T0188425

EDITOR-IN-CHIEF
Douglas Braaten

ASSOCIATE EDITOR
Rebecca E. Cooney

PROJECT MANAGER
Steven E. Bohall

EDITORIAL ADMINISTRATOR
Daniel J. Becker

Artwork and design by Ash Ayman Shairzay

The New York Academy of Sciences
7 World Trade Center
250 Greenwich Street, 40th Floor
New York, NY 10007-2157

annals@nyas.org
www.nyas.org/annals

**The New York
Academy of Sciences**

Published by Blackwell Publishing
On behalf of the New York Academy of Sciences

Boston, Massachusetts
2011

ANNALS *of* THE NEW YORK ACADEMY OF SCIENCES

VOLUME
1240

ISSUE

Skeletal Biology and Medicine II

Bone and cartilage homeostasis and bone disease

ISSUE EDITOR

Mone Zaidi

Mount Sinai School of Medicine

This volume presents manuscripts stemming from the 4th New York Skeletal Biology and Medicine Conference, held at Mount Sinai School of Medicine in New York City on April 27–30, 2011. The papers included in this issue include two of the topic areas presented at the conference; the other topic areas are included in *Skeletal Biology and Medicine I*, Volume 1237.

TABLE OF CONTENTS

Become a Member Today of the New York Academy of Sciences

The New York Academy of Sciences is dedicated to identifying the next frontiers in science and catalyzing key breakthroughs. As has been the case for 200 years, many of the leading scientific minds of our time rely on the Academy for key meetings and publications that serve as the crucial forum for a global community dedicated to scientific innovation.

 Select one FREE *Annals* volume and up to five volumes for only $40 each.

 Network and exchange ideas with the leaders of academia and industry.

 Broaden your knowledge across many disciplines.

 Gain access to exclusive online content.

Join Online at **www.nyas.org**

Or by phone at **800.344.6902** (516.576.2270 if outside the U.S.).

Mechanisms and measurement of bone and cartilage disease

Ann. N.Y. Acad. Sci. ISSN 0077-8923

ANNALS OF THE NEW YORK ACADEMY OF SCIENCES
Issue: *Skeletal Biology and Medicine II*

How are osteoclasts induced to resorb bone?

T.J. Chambers and K. Fuller

Department of Cellular Pathology, St George's University of London, London, United Kingdom

Address for correspondence: T.J. Chambers, Department of Cellular Pathology, St George's University of London, Cranmer Terrace, Tooting, London SW17 0RE, UK. tchamber@sgul.ac.uk

Although much is known about how osteoclasts are formed, we know little about how they are activated, or how they recognize bone as the substrate appropriate for resorption. Bone mineral is considered to be essential to this recognition process, but a "mineral receptor" has never been identified. Recently, we found that resorptive behavior, as judged by the formation of ruffled borders and actin rings, occurs on ordinary tissue culture substrates if they are first coated with vitronectin. Similarly, vitronectin-coated substrates induce osteoclasts to secrete tartrate-resistant acid phosphatase and to form podosome belts, and to make resorption trails in the protein that coat the substrate. The same applies to bone mineral, which only induces resorptive behavior if coated with vitronectin. In contrast, fibronectin has none of these effects, despite inducing adhesion and spreading. It appears that osteoclasts recognize bone as the substrate appropriate for resorption through the high affinity of vitronectin-receptor ligands for bone mineral.

Keywords: bone resorption; osteoclast; podosome belt; vitronectin

When osteoclasts are incubated on a slice of cut bone *in vitro*, they make deep excavations with extraordinary speed. This ability to resorb bone distinguishes them from all other cells. They achieve this by establishing a circle of close contact with the bone surface, associated with the appearance of a peripheral ring of actin, devoid of cytoplasmic organelles (the "clear zone," or "sealing zone"). Vesicles containing proton pumps and acid hydrolases are then inserted into the surface membrane circumscribed within this ring, throwing it into convolutions, called the "ruffled border." Thus, a *resorptive hemivacuole* is formed between cell and bone, in which protons digest the mineral component of bone, and acid hydrolases, predominantly cathepsin K, digest the organic matrix. Dissolved products are transported in vesicles from the resorptive hemivacuole and released at the opposite, basolateral membrane.[1,2] It is highly likely that such behavior is restricted to bone, and it seems to be the mineral component that is responsible. Thus, it has long been known that osteoclasts resorb mineralized, but not unmineralized or demineralized, bone.[3] Our working model has been that resorptive behavior is induced in osteo-clasts by contact with bone mineral, and that this contact occurs when cells of the osteoblastic lineage (bone lining cells and osteocytes) dissolve the layer of unmineralized organic material that normally lines bone surfaces. What is the evidence for this? Bone surfaces are normally covered by a layer of unmineralized organic material,[4,5] and resorption only occurs if the surface organic material is first removed;[3,6,7] *in vivo*, osteoblastic cells in close proximity to osteoclasts express interstitial collagenase;[8] resorptive cytokines and hormones stimulate release of interstitial collagenase by osteoblastic cells;[4] and while both cathepsins and collagenase are essential for resorption in intact bone, resorption of bone slices, in which mineral is exposed by the bone saw, requires only cathepsins.[4,9,10]

Because bone mineral is essential for activation of osteoclasts, a "mineral receptor" has been postulated.[11] However, osteoclasts fail to adhere to bone mineral unless serum is present (unpublished observations). This suggests that an adhesion molecule might also be required, perhaps to facilitate interaction between bone mineral and a putative mineral receptor. A prime candidate for this is the

doi: 10.1111/j.1749-6632.2011.06249.x

vitronectin receptor, $\alpha_v\beta_3$, which is known to be essential for bone resorption: antibodies against $\alpha_v\beta_3$, and $\alpha_v\beta_3$ antagonists, such as echistatin and kistrin, potently inhibit resorption *in vitro* and *in vivo*.[12–16]

We have recently found though that while osteoclasts adhere and spread to a similar degree on bone mineral that has been coated with fibronectin or vitronectin, only vitronectin induces bone resorption.[17] This suggests that the ability of $\alpha_v\beta_3$ to mediate attachment to bone mineral does not explain why it is essential for bone resorption. Perhaps it acts as a cofactor for a mineral receptor in some other way. To analyze this question, we tested the ability of $\alpha_v\beta_3$ ligands to induce resorptive behavior in osteoclasts in the absence of bone mineral.

For these experiments,[17] we exploited a technique we have recently developed whereby osteoclasts can be lifted into suspension and make excavations within three hours of sedimentation onto bone slices.[18] This enabled us to test the response of mature osteoclasts to defined substrates that have not been conditioned by the medium used to generate osteoclasts, or by the osteoclasts themselves. We used these suspensions to test the ability of substrates to induce several markers of resorptive behavior.

First, we tested the ability of substrates to induce secretion of tartrate-resistant acid phosphatase (TRAP), a process that correlates with bone resorption.[19] We observed potent stimulation of TRAP release by osteoclasts incubated on vitronectin-coated glass coverslips. This was abolished by the resorption-inhibiting hormone calcitonin. By contrast, fibronectin not only failed to induce, but actually suppressed TRAP secretion. The differing ability of these vitronectin and fibronectin to activate osteoclasts, despite very similar ability to induce adhesion and spreading in the osteoclasts, was striking. This suggests that vitronectin activates resorptive behavior.

Second, we noted by transmission and scanning electron microscopy that the morphological structures characteristic of resorbing osteoclasts, ruffled borders, and clear zones were formed by osteoclasts incubated on tissue culture substrates coated with vitronectin. Like release of TRAP, these characteristic structures were induced by vitronectin but not by fibronectin.

Third, we noted in the scanning electron microscope that a proteinaceous film could be seen covering the glass substrates. This film was dissolved by osteoclasts incubated on vitronectin, but not fibronectin, to form trails reminiscent of the trails of pits formed by osteoclasts on bone slices *in vitro*. Also similar to bone resorption, the cysteine proteinase inhibitor E64 prevented dissolution of this film, as did resorption-inhibitors, suggesting that dissolution of the protein film was a consequence of resorptive behavior, due to the secretion of cysteine proteinase onto the glass surface. Such trails were not explicable by cell migration alone: they required the presence of resorption-inducing cytokines. These observations strongly suggest that vitronectin-coated substrates induce resorptive behavior in osteoclasts.

Vitronectin coating also induced the formation of podosome belts, although the relationship of these with resorptive behavior has been controversial. On bone, resorbing osteoclasts form a circumferential ring of actin, the "actin ring," corresponding to the sealing zone or clear zone seen in the transmission electron microscope.[20,21] In contrast, osteoclasts on nonmineralized substrates form circumferential belts of podosomes, dot-like structures. The extent to which podosome belts resemble actin rings and signify resorptive behavior is uncertain. It has been claimed that the sealing zone on bone has a different three-dimensional organization that is not derived from podosomes and that podosomes reflect adhesion and migration rather than resorption.[2,11,22] Alternatively, it has been suggested that isolated podosomes fuse to give rise to a continuous sealing zone.[23,24] The latter view is supported by recent evidence that the sealing zone consists of structural units clearly related to individual podosomes, which differ primarily in density and interconnectivity from the podosome belts observed on other substrates.[25]

We found that vitronectin strongly induced the formation of podosome belts on plastic substrates, while fibronectin, which facilitated adhesion and spreading to a similar degree, was unable to induce podosome belt formation.[17] This suggests that podosome belts on tissue culture substrates reflect resorptive behavior rather than cell spreading.

Our experiments[17] provided two lines of evidence that support the notion that podosome belts are the analog on tissue culture surfaces of actin rings on bone. First, we noted that podosome belts correlated more closely with bone resorption than with cell spreading or migration. Thus, podosome belts were stimulated by agents that stimulate bone resorption. By contrast, neither M-CSF, which induces migration and spreading but inhibits bone resorption,[26] nor fibronectin, which facilitated adhesion and spreading but not resorption, increased podosome belt formation.

There is, however, a notable difference between the podosome belts formed on tissue culture substrates and the actin rings formed on bone: their circumference. This difference requires an explanation. Our data[17] suggest that the difference is attributable to the differing topography of the substrates: podosome belts formed on rough perspex were more condensed, and more like the actin rings seen on bone, than the podosome belts formed on smooth perspex. Thus, the lower density of podosome subunits noted on glass compared to mineralized substrates[25] may be due to the greater spreading caused by the smooth surface. An increase in cell spreading would be expected to simultaneously increase podosome belt diameter and decrease podosome density. This observation adds further support to the notion that podosome belts on culture substrates are equivalent to actin rings on bone.

It has been suggested that osteoclasts recognize bone as the substrate appropriate for resorption due to its surface roughness.[27] However, the ability of vitronectin-coated glass coverslips to activate osteoclasts shows that a surface does not necessarily have to be rough to activate resorptive behavior.

Another suggestion has been that osteoclasts detect bone as the resorbogenic substrate by its rigidity. This could explain why demineralized bone induces neither resorption nor podosome belts.[3,28] However, unlike vitronectin, fibronectin coating of rigid substrates does not induce resorptive behavior.[17] Therefore, high substrate rigidity is not sufficient to induce resorptive behavior. Moreover, nonrigid substrates, such as silicone rubber, also induce podosome belts in osteoclasts when coated with vitronectin.[17] A rigid substrate seems to be neither necessary nor sufficient for the induction of resorptive behavior.

In fact, coating surfaces with vitronectin appears to be sufficient to induce all aspects of resorptive behavior in osteoclasts: it induces osteoclasts to form podosome belts, to secrete hydrolytic enzymes, to digest extracellular substrates, and to form clear zones and ruffled borders.

The contrast between the effects of vitronectin and fibronectin on osteoclasts is remarkable. The osteoclast expresses both the vitronectin receptor $\alpha_v\beta_3$[29] and the fibronectin receptor $\alpha_5\beta_1$.[30] Both vitronectin and fibronectin enable osteoclasts to attach to glass coverslips or bone mineral in the absence of serum, and both facilitate cell spreading.

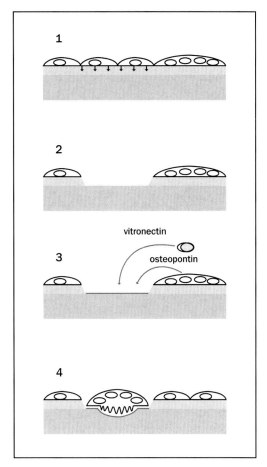

Figure 1. Model for the mechanism of activation of bone resorption. (1) When osteoblastic cells determine the site appropriate for resorption, they secrete interstitial collagenase, which digests the unmineralized organic material that coats the surfaces of (mineralized) bone. (2) This exposes onto the surface $\alpha_v\beta_3$ ligands that were adsorbed onto bone mineral during bone formation. (3) The freshly exposed mineralized surface may well adsorb additional $\alpha_v\beta_3$ ligands from serum (vitronectin) and osteoclasts (osteopontin, bone sialoprotein), especially as many such ligands possess a multiphosphorylated motif with a strong affinity for bone mineral. (4) Thus, bone is recognized not directly but through the high affinity of bone mineral for $\alpha_v\beta_3$ ligands, which are both necessary and sufficient to activate resorptive behavior in the osteoclast.

Yet only vitronectin induces resorptive behavior.[17] c-Src is essential for bone resorption,[31–33] so that the likely molecular basis for this observation is that c-Src binds selectively to β_3 not β_1 integrins, and that clustering of β_3 activates c-Src;[34] c-Src deficiency has no detectable effects on fibronectin-receptor function.[35]

Vitronectin receptor ligands such as osteopontin and thrombospondin are highly expressed by bone cells and have a strong affinity for bone mineral through a multiphosphorylated motif.[36–39] Such ligands, incorporated in bone during bone formation, might activate osteoclasts when bone mineral is exposed by osteoblastic cells.[6] Alternatively or additionally, activation might follow binding of osteopontin or bone sialoprotein, which are known to be expressed by osteoclasts,[40–43] to freshly exposed bone mineral that or binding of $\alpha_v\beta_3$ ligands secreted by osteoclast-regulatory osteoblastic cells independently of bone formation, or vitronectin from serum, in which it is present at between 300 μg/mL and 700 μg/mL.[44,45]

Thus, it appears that it is $\alpha_v\beta_3$ ligands, rather than bone mineral, that are necessary and sufficient for the induction of resorptive behavior in osteoclasts; bone mineral is not recognized directly, though it possesses a high affinity for $\alpha_v\beta_3$ ligands. This provides a model through which osteoblastic cells can extend their ability to regulate osteoclasts to include the induction and localization of resorption, by exposing mineral onto the bone surface through secretion of interstitial collagenase.[4] Once exposed, bone mineral, by binding endogenous or exogenous $\alpha_v\beta_3$ ligands, activates resorptive behavior in the osteoclast (Fig. 1).

However, this apparently unique ability of $\alpha_v\beta_3$ ligands to activate resorption is at odds with the recent conclusion, using mice deleted for one or more integrin subunits, that multiple integrin classes are required for osteoclast-mediated bone resorption.[46] Furthermore, although antibody and inhibitor studies show convincingly that $\alpha_v\beta_3$ ligands are essential for bone resorption, mice deleted for the β_3 component of $\alpha_v\beta_3$ do not have the osteopetrosis we would expect to see in animals with severely defective osteoclasts.[46,47]

The most likely explanation is redundancy, but which receptor system is responsible? Osteoclasts express not only $\alpha_v\beta_3$ and $\alpha_5\beta_1$, but $\alpha_2\beta_1$, the collagen receptor.[48] It has been suggested that $\alpha_2\beta_1$ can

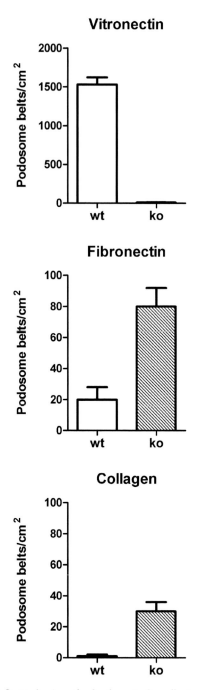

Figure 2. Mechanism of redundancy in β_3-null mice. Osteoclasts were formed from the bone marrow of β_3-null (ko) or wild-type (wt) mice. They were then lifted into suspension and sedimented onto glass coverslips previously coated with vitronectin, fibronectin, or collagen (50 μg/mL). Cells were incubated for five hours before staining for F-actin to enable counting of podosome belts. Osteoclasts from both sources formed a confluent monolayer on all three substrates. Results are derived from three cultures per variable.

substitute for loss of β_3.[49] Osteoclasts express the components of $\alpha_v\beta_1$, which can act as a vitronectin receptor.[50] We tested the nature of redundancy by sedimenting mature bone marrow-derived osteoclasts, as described previously,[17] from mice deleted for the β_3 subunit, onto glass coverslips coated with adhesion factors. We found (Fig. 2) that while osteoclasts formed from β_3-null mice produced no podosome belts on vitronectin-coated coverslips, they formed substantially greater numbers than did wild-type osteoclasts when incubated on coverslips coated with fibronectin or collagen. Thus, there is partial compensation, not by an alternative vitronectin receptor, but by fibronectin and collagen receptors. This partial compensation is likely to be augmented by the substantially greater numbers of osteoclasts on bone surfaces in β_3-null mice.[47]

The mechanistic basis for compensation might be that signaling molecules normally involved in β_3 signaling become available, in β_3-null mice, for fibronectin and collagen receptors. Alternatively or additionally, expression of the latter receptors might increase. In either case, our results suggest that the ability to induce bone resorption is not necessarily specific to $\alpha_v\beta_3$: in β_3-null osteoclasts fibronectin, which can clearly bind to bone mineral, and collagen, which is present in bone matrix in huge quantities, are available as alternative ligands. It appears though that these ligands, and their receptors, do not play a significant role in normal osteoclasts, in which it is $\alpha_v\beta_3$, as the dominant signaling system, that is responsible for the recognition of bone as the substrate appropriate for resorption through the ability of bone mineral to bind $\alpha_v\beta_3$ substrates.

Conflicts of interest

The authors declare no conflicts of interest.

References

1. Stenbeck, G. 2002. Formation and function of the ruffled border in osteoclasts. *Semin. Cell. Dev. Biol.* **13:** 285–292.
2. Vaananen, H.K., H. Zhao, M. Mulari & J.M. Halleen. 2000. The cell biology of osteoclast function. *J. Cell. Sci.* **113:** 377–381.
3. Chambers, T.J., B.M. Thomson & K. Fuller. 1984. Effect of substrate composition on bone resorption by rabbit osteoclasts. *J. Cell Sci.* **70:** 61–71.
4. Chambers, T.J. & T.J. Hall. 1991. Cellular and molecular mechanisms in the regulation and function of osteoclasts. *Vitam. Horm.* **46:** 41–86.
5. Chow, J. & T.J. Chambers. 1992. An assessment of the preva-
lence of organic material on bone surfaces. *Calcif. Tissue Int.* **50:** 118–122.
6. Chambers, T.J. & K. Fuller. 1985. Bone cells predispose bone surfaces to resorption by exposure of mineral to osteoclastic contact. *J. Cell Sci.* **76:** 155–165.
7. Chambers, T.J., J.A. Darby & K. Fuller. 1985. Mammalian collagenase predisposes bone surfaces to osteoclastic resorption. *Cell Tissue Res.* **241:** 671–675.
8. Fuller, K. & T.J. Chambers. 1995. Localisation of mRNA for collagenase in osteocytic, bone surface and chondrocytic cells but not osteoclasts. *J. Cell Sci.* **108:** 2221–2230.
9. Delaisse, J.M., A. Boyde, E. Maconnachie, *et al.* 1987. The effects of inhibitors of cysteine-proteinases and collagenase on the resorptive activity of isolated osteoclasts. *Bone* **8:** 305–313.
10. Fuller, K., B. Kirstein & T.J. Chambers. 2007. Regulation and enzymatic basis of bone resorption by human osteoclasts. *Clin. Sci. (London)* **112:** 567–575.
11. Saltel, F., O. Destaing, F. Bard, *et al.* 2004. Apatite-mediated actin dynamics in resorbing osteoclasts. *Mol. Biol. Cell* **15:** 5231–5241.
12. Fisher, J.E., M.P. Caulfield, M. Sato, *et al.* 1993. Inhibition of osteoclastic bone-resorption in vivo by echistatin, an arginyl-glycyl-aspartyl (rgd)-containing protein. *Endocrinology* **132:** 1411–1413.
13. Lakkakorpi, P.T., M.A. Horton, M.H. Helfrich, *et al.* 1991. Vitronectin receptor has a role in bone-resorption but does not mediate tight sealing zone attachment of osteoclasts to the bone surface. *J. Cell Biol.* **115:** 1179–1186.
14. Engelman, V.W., G.A. Nickols, F.P. Ross, *et al.* 1997. A peptidomimetic antagonist of the alpha(v)beta(3) integrin inhibits bone resorption in vitro and prevents osteoporosis in vivo. *J. Clin. Invest.* **99:** 2284–2292.
15. Chambers, T.J., K. Fuller, J.A. Darby, *et al.* 1986. Monoclonal antibodies against osteoclasts inhibit bone resorption in vitro. *Bone Miner.* **1:** 127–135.
16. Horton, M.A., M.L. Taylor, T.R. Arnett & M.H. Helfrich. 1991. Arg-Gly-Asp (RGD) peptides and the anti-vitronectin receptor antibody 23C6 inhibit dentine resorption and cell spreading by osteoclasts. *Exp. Cell Res.* **195:** 368–375.
17. Fuller, K., J.L. Ross, K.A. Szewczyk, *et al.* 2010. Bone is not essential for osteoclast activation. *PLoS One* **5:** e12837.
18. Fuller, K., B. Kirstein & T.J. Chambers. 2006. Murine osteoclast formation and function: differential regulation by humoral agents. *Endocrinology* **147:** 1979–1985.
19. Kirstein, B., T.J. Chambers & K. Fuller. 2006. Secretion of tartrate-resistant acid phosphatase by osteoclasts correlates with resorptive behavior. *J. Cell Biochem.* **98:** 1085–1094.
20. Lakkakorpi, P.T. & H.K. Vaananen. 1990. Calcitonin, prostaglandin E2, and dibutyryl cyclic adenosine 3',5'-monophosphate disperse the specific microfilament structure in resorbing osteoclasts. *J. Histochem. Cytochem.* **38:** 1487–1493.
21. Burgess, T.L., Y. Qian, S. Kaufman, *et al.* 1999. The ligand for osteoprotegerin (OPGL) directly activates mature osteoclasts. *J. Cell Biol.* **145:** 527–538.
22. Jurdic, P., F. Saltel, A. Chabadel & O. Destaing. 2006. Podosome and sealing zone: specificity of the osteoclast model. *Eur. J. Cell Biol.* **85:** 195–202.

23. Lakkakorpi, P.T., M.H. Helfrich, M.A. Horton & H.K. Vaananen. 1993. Spatial organization of microfilaments and vitronectin receptor, alpha v beta 3, in osteoclasts. A study using confocal laser scanning microscopy. *J. Cell Sci.* **104:** 663–670.

24. Lakkakorpi, P.T. & H.K. Vaananen. 1996. Cytoskeletal changes in osteoclasts during the resorption cycle. *Microsc. Res. Tech.* **33:** 171–181.

25. Luxenburg, C., D. Geblinger, E. Klein, *et al.* 2007. The architecture of the adhesive apparatus of cultured osteoclasts: from podosome formation to sealing zone assembly. *PLoS One* **2:** e179.

26. Fuller, K., J.M. Owens, C.J. Jagger, *et al.* 1993. Macrophage colony-stimulating factor stimulates survival and chemotactic behavior in isolated osteoclasts. *J. Exp. Med.* **178:** 1733–1744.

27. Geblinger, D., L. Addadi & B. Geiger. 2010. Nano-topograpy sensing by osteoclasts. *J. Cell Sci.* **123:** 1503–1510.

28. Nakamura, I., N. Takahashi, T. Sasaki, *et al.* 1996. Chemical and physical properties of the extracellular matrix are required for the actin ring formation in osteoclasts. *J. Bone Miner. Res.* **11:** 1873–1879.

29. Charo, I.F., L. Nannizzi, J.W. Smith & D.A. Cheresh. 1990. The vitronectin receptor alpha v beta 3 binds fibronectin and acts in concert with alpha 5 beta 1 in promoting cellular attachment and spreading on fibronectin. *J. Cell Biol.* **111:** 2795–2800.

30. Hughes, D.E., D.M. Salter, S. Dedhar & R. Simpson. 1993. Integrin expression in human bone. *J. Bone Miner. Res.* **8:** 527–533.

31. Lowe, C., T. Yoneda, B.F. Boyce, *et al.* 1993. Osteopetrosis in src-deficient mice is due to an autonomous defect of osteoclasts. *Proc. Natl. Acad. Sci. U. S. A.* **90:** 4485–4489.

32. Boyce, B.F., T. Yoneda, C. Lowe, *et al.* 1992. Requirement of pp60c-src expression for osteoclasts to form ruffled borders and resorb bone in mice. *J. Clin. Invest.* **90:** 1622–1627.

33. Horne, W.C., L. Neff, D. Chatterjee, *et al.* 1992. Osteoclasts express high-levels of pp60(c-src) in association with intracellular membranes. *J. Cell Biol.* **119:** 1003–1013.

34. Arias-Salgado, E.G., S. Lizano, S. Sarkar, *et al.* 2003. SRC kinase activation by direct interaction with the integrin beta cytoplasmic domain. *Proc. Natl. Acad. Sci. U. S. A.* **100:** 13298–13302.

35. Felsenfeld, D.P., P.L. Schwartzberg, A. Venegas, *et al.* 1999. Selective regulation of integrin-cytoskeleton interactions by the tyrosine kinase SRC. *Nature Cell Biol.* **1:** 200–206.

36. Boskey, A.L., M. Maresca, W. Ullrich, *et al.* 1993. Osteopontin–hydroxyapatite interactions in-vitro: inhibition of hydroxyapatite formation and growth in a gelatin-gel. *Bone Miner.* **22:** 147–159.

37. Oldberg, A., A. Franzen & D. Heinegard. 1986. Cloning and sequence-analysis of rat bone sialoprotein (osteopontin) CDNA reveals an Arg-Gly-Asp cell-binding sequence. *Proc. Natl. Acad. Sci. U. S. A.* **83:** 8819–8823.

38. Huq, N.L., K.J. Cross & E.C. Reynolds. 2000. Molecular modelling of a multiphosphorylated sequence motif bound to hydroxyapatite surfaces. *J. Mol. Modeling* **6:** 35–47.

39. Robey, P.G., M.F. Young, L.W. Fisher & T.D. McClain. 1989. Thrombospondin is an osteoblast-derived component of mineralized extracellular matrix. *J. Cell Biol.* **108:** 719–727.

40. Bianco, P., L.W. Fisher, M.F. Young, *et al.* 1991. Expression of bone sialoprotein (BSP) in developing human tissues. *Calc. Tiss. Intl.* **49:** 421–426.

41. Dodds, R.A., J.R. Connor, I.E. James, *et al.* 1995. Human osteoclasts, not osteoblasts, deposit osteopontin onto resorption surfaces: an in vitro and ex vivo study of remodeling bone. *J. Bone Miner. Res.* **10:** 1666–1680.

42. Arai, N., K. Ohya, S. Kasugai, *et al.* 1995. Expression of bone sialoprotein messenger-RNA during bone-formation and resorption induced by colchicine in rat tibial bone-marrow cavity. *J. Bone Miner. Res.* **10:** 1209–1217.

43. Tezuka, K., T. Sato, H. Kamioka, *et al.* 1992. Identification of osteopontin in isolated rabbit osteoclasts. *Biochem. Biophys. Res. Comm.* **186:** 911–917.

44. Boyd, N.A., A.R. Bradwell & R.A. Thompson. 1993. Quantitation of vitronectin in serum: evaluation of its usefulness in routine clinical practice. *J. Clin. Pathol.* **46:** 1042–1045.

45. Shaffer, M.C., T.P. Foley & D.W. Barnes. 1984. Quantitation of spreading factor in human biologic fluids. *J. Lab. Clin. Med.* **103:** 783–791.

46. Schmidt, S., I. Nakchbandi, R. Ruppert, *et al.* 2011. Kindlin-3-mediated signaling from multiple integrin classes is required for osteoclast-mediated bone resorption. *J. Cell Biol.* **192:** 883–897.

47. McHugh, K.P., K. Hodivala-Dilke, M.H. Zheng, *et al.* 2000. Mice lacking beta3 integrins are osteosclerotic because of dysfunctional osteoclasts. *J. Clin. Invest.* **105:** 433–440.

48. Helfrich, M.H., S.A. Nesbitt, P.T. Lakkakorpi, *et al.* 1996. Beta 1 integrins and osteoclast function: involvement in collagen recognition and bone resorption. *Bone* **19:** 317–328.

49. Horton, M.A., H.M. Massey, N. Rosenberg, *et al.* 2003. Upregulation of osteoclast alpha2beta1 integrin compensates for lack of alphavbeta3 vitronectin receptor in Iraqi-Jewish-type Glanzmann thrombasthenia. *Br. J. Haematol.* **122:** 950–957.

50. Koistinen, P. & J. Heino. 2002. The selective regulation of alpha Vbeta 1 integrin expression based on the hierarchical formation of alpha V-containing heterodimers. *J. Biol. Chem.* **277:** 24835–24841.

Ann. N.Y. Acad. Sci. ISSN 0077-8923

ANNALS OF THE NEW YORK ACADEMY OF SCIENCES
Issue: *Skeletal Biology and Medicine II*

Epigenetic regulation of osteoclast differentiation

Tetsuro Yasui,[1] Jun Hirose,[1] Hiroyuki Aburatani,[2] and Sakae Tanaka[1]

[1]Department of Orthopaedic Surgery, The University of Tokyo, Tokyo, Japan. [2]Genome Science Division, Research Center for Advanced Science and Technology, The University of Tokyo, Tokyo, Japan

Address for correspondence: Sakae Tanaka, Department of Orthopaedic Surgery, Faculty of Medicine, The University of Tokyo, Hongo 7-3-1, Bunkyo-ku, Tokyo 113-0033, Japan. TANAKAS-ORT@h.u-tokyo.ac.jp

Recent studies have uncovered that epigenetic regulation, such as histone methylation and acetylation, plays a critical role in determining cell fate. In particular, the expression of key developmental genes tends to be regulated by trimethylation of histone H3 lysine 4 (H3K4me3) and lysine 27 (H3K27me3). Osteoclasts are primary cells for bone resorption, and their differentiation is tightly regulated by the receptor activator of nuclear factor κB ligand (RANKL) and a transcription factor nuclear factor–activated T cell (NFAT) c1. We found that RANKL-induced NFATc1 expression is associated with the demethylation of H3K27me3. Jumonji domain containing-3, a H3K27 demethylase, is induced in bone marrow–derived macrophages in response to RANKL stimulation and may play a critical role in the demethylation of H3K27me3 in the *Nfatc1* gene.

Keywords: epigenetics; osteoclast; differentiation; histone modification; Jmjd3

Epigenetic regulation of cell differentiation

All cells in the animal body share an identical genome, but there are many types of cells with different behaviors. There must be a tight and concrete mechanism of lineage commitment when cells progress from totipotency to pluripotency to terminal differentiation. This has been one of the fundamental mysteries of biology; however, accumulating knowledge of epigenetic mechanisms is now unveiling it.

Epigenetics is defined as heritable changes in the function of genetic elements without changes in the DNA sequence.[1] Different from the genome, the epigenome is dynamic and changes in response to the environment and the situation. There are three classes of epigenetic marks: DNA methylation, histone modification, and noncoding RNAs (Table 1). These marks solely or cooperatively tune gene expression profiles positively or negatively (Fig. 1). Recently established genome-wide analytic methods, such as chromatin immunoprecipitation (ChIP)-chip, ChIP-sequencing (ChIP-seq), and bisulfate sequencing, have enabled us to construct a systemic map of human and mouse epigenome, which allows us a better understanding of the epigenetic regulation of cells.[2,3]

DNA methylation

DNA methylation has been the most intensively studied type of epigenetic modification since its discovery in the late 1940s.[4] DNA methylation of mammalian cells is predominantly observed as 5-methylcytosine on the pyrimidine ring at the cytosine residue of CpG dinucleotides.[5] Hypermethylation of CpG-rich regions (CpG islands) in gene promoters generally blocks the gene expression, and hypomethylation leads to active transcription.

Methylation of CpG islands has been analyzed in the context of cancer because hypermethylation of tumor suppressor genes and hypomethylation of oncogenes have frequently been observed in cancer cells.[6,7] However, recent studies have revealed that DNA methylation of surrounding regions of CpG islands ("CpG island shores") regulates the expression of tissue-specific genes, and thus contributes to the lineage commitment of each cell during differentiation.[8,9] For example, Ji *et al.* clearly depicted that the CpG island shore of lymphocyte-specific protein tyrosine kinase (*Lck*), which is responsible in determination of cell fate toward T cells,[10] is demethylated during differentiation from multipotent progenitors to thymic T cell progenitors.[9]

doi: 10.1111/j.1749-6632.2011.06245.x

Table 1. Typical classes of epigenetic regulation

- DNA methylation
  DNA methylation of CpG island
  DNA methylation of CpG island shore
- Histone modification
  Histone acetylation
  Histone methylation
  Histone phosphorylation
  Histone ubiquitylation
- Noncoding RNAs
  microRNAs (miRNAs)
  small nucleolar RNAs (snoRNAs)
  small interfering RNAs (siRNAs)
  PIWI-interacting RNAs (piRNAs)
  transcription initiation RNAs (tiRNAs)
  enhancer RNAs (eRNAs)
  long noncoding RNAs (lncRNAs)

Histone modification

In mammalian cells, the basic building block of chromatin is the nucleosome, which is composed of approximately 146 bp of DNA wrapped around a protein octamer consisting of two copies of four dis-

tinct histone proteins (H2A, H2B, H3, and H4).[11] N-terminal tails of histones are subject to several types of modification, including acetylation, methylation, phosphorylation, and ubiquitylation.[12] These modifications regulate the accessibility of transcription factors and cofactors to specific sites of the genome, and thus control the activation or repression of associated genes. Acetylation of histone by histone acetyltransferases (HATs) leads to active transcription of the gene, and deacetylation of histone by histone deacetylases (HDACs) leads to gene silencing.[13] Histone methylation plays a pivotal role in both positive and negative regulations of gene expression, depending on the sites of methylation, and has been shown to regulate normal development, carcinogenesis, and immune responses.[14,15] Trimethylation of histone H3 at lysine-4, -36, and -79 (H3K4, H3K36, and H3K79) is implicated in the activation of transcription, whereas trimethylation of histone H3 at lysine-9 and -27, and histone H4 at lysine-20 (H3K9, H3K27, and H4K20) is associated with the repression of transcription.[16,17]

Each modification alone affects gene transcription, but the combination of these modifications determines the distinct chromatin state and governs

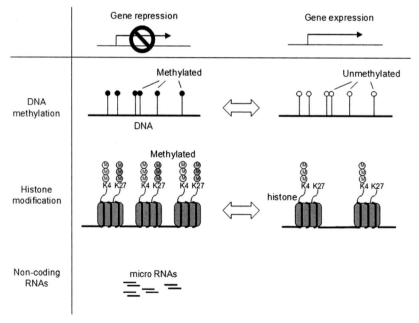

Figure 1. Typical epigenetic marks and gene expression. There are three classes of epigenetic marks: DNA methylation, histone modification, and noncoding RNAs. Epigenetic marks, such as DNA methylation, trimethylation of lysine 27 in histone H3 (H3K27), or microRNAs, negatively regulate gene expression, whereas DNA demethylation, trimethylation of lysine 4 in histone H3, or demethylation of H3K27 positively regulates it.

gene transcription.[18] Some promoters in embryonic stem cells are known to carry both trimethylation of H3K4 (H3K4me3) and trimethylation of H3K27 (H3K27me3).[19,20] Bernstein *et al.* termed this combination a "bivalent" chromatin mark and reported that it serves to poise key developmental genes for lineage-specific activation or repression.[20] The trithorax group proteins catalyze H3K4me3, and polycomb group proteins catalyze H3K27me3. Histone demethylases that specifically demethylate H3K27me3, such as Jmjd3 and Utx, were recently identified, and the involvement of these molecules in various physiologic and pathologic conditions has been reported.[21–24] For example, using Jmjd3 knock-out mice, Satoh *et al.* reported that Jmjd3 regulates M2 macrophage polarization and the host response to helminth infection.[23]

Noncoding RNAs

Noncoding RNAs are functional RNAs without being translated into proteins and include not only transfer RNA and ribosomal RNA but also recently classified small nucleolar RNAs (snoRNAs), microRNAs (miRNAs), small interfering RNAs (siRNAs), PIWI-interacting RNAs (piRNAs), transcription initiation RNAs (tiRNAs), enhancer RNAs (eRNAs), and long noncoding RNAs.[25–28] The entire picture of the diversity and functions of noncoding RNAs is still unclear, but recent studies have been unveiling their important roles in the epigenetic regulation of cells.

One major class of noncoding RNAs are miRNAs, which inhibit gene expression by hybridizing to target sites with complementary sequences.[29] Mature miRNAs are cleaved from ~70-nucleotide hairpin structures called precursor miRNAs (pre-miRNAs) by the enzyme Dicer; pre-miRNAs are, in turn, excised from a primary miRNA (pri-mRNA) transcript by the enzyme Drosha.[30]

As an example of miRNA-related regulation of cell differentiation, Chen *et al.* reported that miRNA-181, which is highly expressed in B lymphoid cells and potentially represses myeloid cell leukemia sequence 1 (*Mcl1*) transcription, regulates lineage commitment for B lymphoid cells.[31,32]

Epigenetic regulation of osteoclast differentiation

Osteoclasts are terminally differentiated multinuclear giant cells that are primarily responsible for bone resorption.[33–35] Abnormal differentiation or function in the cells results in unwanted skeletal homeostasis and causes diseases such as osteoporosis and osteopetrosis.[36]

Two key cytokines, macrophage colony-stimulating factor (M-CSF) and the receptor activator of nuclear factor κB ligand (RANKL), mainly provided by stromal/osteoblastic cells, play essential roles in osteoclast differentiation from hematopoietic stem cells.[37,38]

Transcription factors such as microphthalmia transcription factor (MITF), c-Fos, NF-κB, and nuclear factor–activated T cells (NFATc1) are known as important players in the regulation of osteoclast differentiation. Among them, NFATc1 is a master regulator which is indispensable for osteoclast differentiation.[39,40] The expression of NFATc1 in M-CSF-dependent bone marrow-derived macrophages (BMMs) is upregulated by RANKL stimulation, and NFATc1 subsequently induces a series of osteoclast-specific genes, including cathepsin K, tartrate-resistant acid phosphatase (TRAP), calcitonin receptor, vacuolar type ATPase, osteoclast-associated receptor (OSCAR), and β3-integrin, in cooperation with other aforementioned transcription factors.

Several studies have revealed that the epigenetic regulatory mechanisms are involved in osteoclast differentiation. Regarding DNA methylation, Kitazawa and Kitazawa reported that the expression of the *rankl* gene in murine bone marrow stromal cells is epigenetically regulated by DNA methylation of CpG islands around the transcription start sites.[41]

As for miRNAs, indispensability in osteoclast differentiation was shown by the finding that knockdown of Dicer in myeloid-osteoclast lineage cells attenuated osteoclast formation.[42,43] There are also several reports regarding the function of specific miRNAs in osteoclast differentiation. Sugatani *et al.* reported the importance of miRNA-223 in osteoclast differentiation.[43] miRNA-223 mediates differentiation of osteoclasts probably through negative regulation of nuclear factor I-A (NFI-A), reportedly a key factor for the expression of M-CSF receptor. The same group underwent microarray analysis and identified miRNA-21 as an important factor in osteoclastogenesis.[44] They demonstrated that the downregulation of miRNA-21 leads to impaired osteoclast differentiation through the upregulation of programmed cell death 4 (PDCD4), a repressor of c-Fos. miRNA as a repressor for osteoclast lineage commitment has also been reported. Mann *et al.* recently reported a unique role of miRNA-155 in lineage commitment of the monocyte lineage cells for macrophage differentiation and repressing osteoclast formation via repressive transcription of MITF.[45]

Very little is known about the regulation of osteoclast differentiation through histone modification, and we recently proposed a regulatory mechanism of osteoclast differentiation through H3K27 demethylation of Nfatc1.[46]

Possible involvement of Jmjd3 in the histone demethylation of Nfatc1 during osteoclastogenesis

As mentioned previously, recent studies have revealed the importance of histone methylation patterns of H3K4 and H3K27 in lineage commitment during cell differentiation. We recently reported that a key transcription factor of osteoclast differentiation, NFATc1, is epigenetically regulated by demethylation of H3K27me3 by Jmjd3.

By employing ChIP-seq technology, we analyzed the H3K4me3 and H3K27me3 modification patterns around the transcription start sites (TSSs) of several transcription factors important for osteoclastogenesis, i.e. *Mitf*, *Nfkb1*, *Nfkb2*, *Fos*, and *Nfatc1*. As shown in Figure 2, H3K4me3 marks were present in both BMMs and osteoclasts in all of these transcription factors except for *Mitf*. The H3K27me3 marks were present in the TSS of *Nfatc1*, but not of the other transcription factors. Following the treatment with RANKL and subsequent osteoclast differentiation, a marked reduction in the level of H3K27me3 at the locus was observed (Fig. 2). These results demonstrated that RANKL-induced osteoclast differentiation is associated with a dynamic change in the histone modification from bivalent H3K4me3/H3K27me3 to monovalent H3K4me3. Through ChIP-PCR (polymerase chain reaction) analysis, we identified that Jmjd3, demethylase of H3K27me3, is recruited in the vicinity of the TSS of *Nfatc1* in response to RANKL stimulation (Fig. 3A). Knockdown of the *Jmjd3* gene by shRNA attenuated the demethylation of H3K27me3 at the Nfatc1 locus and suppressed RANKL-induced Nfatc1 induction and osteoclast differentiation (Figs 3B–E). These results suggest that demethylation of H3K27me3 in the vicinity of the TSS of the *Nfatc1*, regulated by Jmjd3, plays a key role in RANKL-induced osteoclast differentiation.

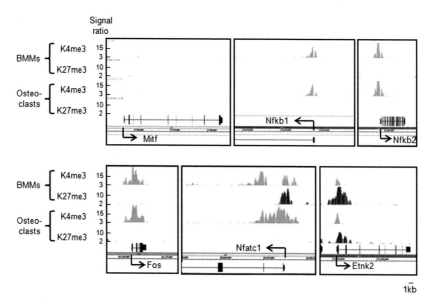

Figure 2. Dynamic change in the histone modification pattern in transcription factors important for osteoclast differentiation (adapted from Ref. 43, Fig. 2). Global information on the H3K4me3 and H3K27me3 modification profiles in BMMs and osteoclasts (OCs) were obtained by ChIP-seq analysis, and data from *Mitf*, *Nfkb1*, *Nfkb2*, *Fos*, and *Nfatc1* were extracted. The data from *Etnk2* serves as a positive control for both H3K4me3 and H3K27me3. Note that H3K27me3 was markedly reduced in *Nfatc1* in osteoclasts.

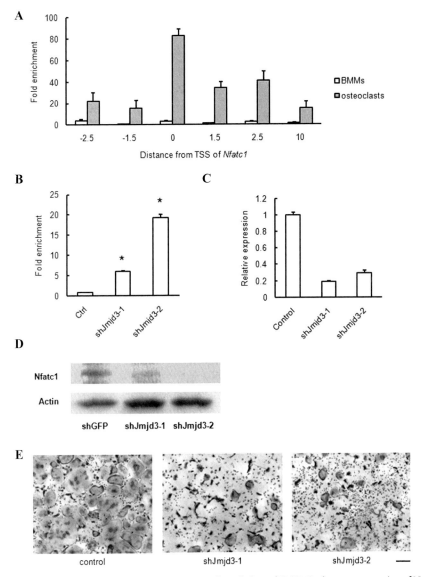

Figure 3. Essential role of Jmjd3 in osteoclast differentiation as a demethylase of H3K27 in the promoter region of *Nfatc1* (modified from Ref. 43). (A) Chromatin solution from BMMs and osteoclasts was subjected to ChIP analysis using an anti-Jmjd3 antibody. The obtained DNA was analyzed with real-time PCR with primer pairs amplifying the several promoter region of *Nfatc1*. Fold enrichment was calculated using the promoter region of *Pparg2* as a negative control. Note that Jmjd3 was recruited around the TSS of *Nfatc1* by stimulation with RANKL. (B–E) BMMs transfected with shJmjd3 were stimulated with 5 nM GST-RANKL (kindly provided by Oriental Yeast Co. Ltd.) for 72 hours. (B) Trimethylation of H3K27 in the promoter region of the *Nfatc1* gene was assessed by ChIP-PCR. *P < 0.05 versus control. (C) mRNA expression of *Jmjd3* was assessed with real-time PCR. (D) Expression of NFATc1 was assessed with Western blot. (E) TRAP staining. Bar = 100 μm. Note that knockdown of Jmjd3 suppressed RANKL-induced H3K27 demethylation in the TSS of *Nfatc1*, RANKL-induced expression of NFATc1, and RANKL-induced osteoclastogenesis.

Conclusion

Epigenetic regulation of cell differentiation is one of the major research topics of biology; we have, however, been gaining a better understanding of this mechanism through accumulating work. Further progress in exploring the epigenetic regulation of cell differentiation is expected through recently established genome-wide methods and more comprehensive analyses.

Conflicts of interest

The authors declare no conflicts of interest.

References

1. Bird, A. 2007. Perceptions of epigenetics. *Nature* **447:** 396–398.
2. Barski, A. *et al.* 2007. High-resolution profiling of histone methylations in the human genome. *Cell* **129:** 823–837.
3. Cokus, S.J. *et al.* 2008. Shotgun bisulphite sequencing of the Arabidopsis genome reveals DNA methylation patterning. *Nature* **452:** 215–219.
4. Hotchkiss, R.D. 1948. The quantitative separation of purines, pyrimidines, and nucleosides by paper chromatography. *J. Biol. Chem.* **175:** 315–332.
5. Jaenisch, R. & A. Bird. 2003. Epigenetic regulation of gene expression: how the genome integrates intrinsic and environmental signals. *Nat. Genet.* **33**(Suppl): 245–254.
6. Feinberg, A.P. 2007. Phenotypic plasticity and the epigenetics of human disease. *Nature* **447:** 433–440.
7. Feinberg, A.P. & B. Tycko. 2004. The history of cancer epigenetics. *Nat. Rev. Cancer* **4:** 143–153.
8. Doi, A. *et al.* 2009. Differential methylation of tissue- and cancer-specific CpG island shores distinguishes human induced pluripotent stem cells, embryonic stem cells and fibroblasts. *Nat. Genet.* **41:** 1350–1353.
9. Ji, H. *et al.* 2010. Comprehensive methylome map of lineage commitment from haematopoietic progenitors. *Nature* **467:** 338–342.
10. Molina, T.J. *et al.* 1992. Profound block in thymocyte development in mice lacking p56lck. *Nature* **357:** 161–164.
11. Luger, K. *et al.* 1997. Crystal structure of the nucleosome core particle at 2.8 A resolution. *Nature* **389:** 251–260.
12. Jenuwein, T. & C.D. Allis. 2001. Translating the histone code. *Science* **293:** 1074–1080.
13. Zardo, G., G. Cimino & C. Nervi. 2008. Epigenetic plasticity of chromatin in embryonic and hematopoietic stem/progenitor cells: therapeutic potential of cell reprogramming. *Leukemia* **22:** 1503–1518.
14. Esteller, M. 2008. Epigenetics in cancer. *N. Eng. J. Med.* **358:** 1148–1159.
15. Roh, T.Y. *et al.* 2006. The genomic landscape of histone modifications in human T cells. *Proc. Natl. Acad. Sci. USA* **103:** 15782–15787.
16. Cui, K. *et al.* 2009. Chromatin signatures in multipotent human hematopoietic stem cells indicate the fate of bivalent genes during differentiation. *Cell Stem Cell* **4:** 80–93.
17. Yu, H. *et al.* 2008. Inferring causal relationships among different histone modifications and gene expression. *Genome Res.* **18:** 1314–1324.
18. Strahl, B.D. & C.D. Allis. 2000. The language of covalent histone modifications. *Nature* **403:** 41–45.
19. Azuara, V. *et al.* 2006. Chromatin signatures of pluripotent cell lines. *Nat. Cell Biol.* **8:** 532–538.
20. Bernstein, B.E. *et al.* 2006. A bivalent chromatin structure marks key developmental genes in embryonic stem cells. *Cell* **125:** 315–326.
21. Agger, K. *et al.* 2007. UTX and JMJD3 are histone H3K27 demethylases involved in HOX gene regulation and development. *Nature* **449:** 731–734.
22. Hong, S. *et al.* 2007. Identification of JmjC domain-containing UTX and JMJD3 as histone H3 lysine 27 demethylases. *Proc. Natl. Acad. Sci. USA* **104:** 18439–18444.
23. Satoh, T. *et al.* 2010. The Jmjd3-Irf4 axis regulates M2 macrophage polarization and host responses against helminth infection. *Nat. Immunol.* **11:** 936–944.
24. Sen, G.L. *et al.* 2008. Control of differentiation in a self-renewing mammalian tissue by the histone demethylase JMJD3. *Genes Dev.* **22:** 1865–1870.
25. Kim, T.K. *et al.* 2010. Widespread transcription at neuronal activity-regulated enhancers. *Nature* **465:** 182–187.
26. Morris, K.V. 2009. Non-coding RNAs, epigenetic memory and the passage of information to progeny. *RNA Biol.* **6:** 242–247.
27. Taft, R.J. *et al.* 2009. Tiny RNAs associated with transcription start sites in animals. *Nat. Genet.* **41:** 572–578.
28. Ponting, C.P., P.L. Oliver & W. Reik. 2009. Evolution and functions of long noncoding RNAs. *Cell* **136:** 629–641.
29. Bartel, D.P. & C.Z. Chen. 2004. Micromanagers of gene expression: the potentially widespread influence of metazoan microRNAs. *Nat. Rev. Genet.* **5:** 396–400.
30. Chen, K. & N. Rajewsky. 2007. The evolution of gene regulation by transcription factors and microRNAs. *Nat. Rev. Genet.* **8:** 93–103.
31. Chen, C.Z. *et al.* 2004. MicroRNAs modulate hematopoietic lineage differentiation. *Science* **303:** 83–86.
32. Zimmerman, E.I. *et al.* 2010. Lyn kinase-dependent regulation of miR181 and myeloid cell leukemia-1 expression: implications for drug resistance in myelogenous leukemia. *Mol. Pharmacol.* **78:** 811–817.
33. Boyle, W.J., W.S. Simonet & D.L. Lacey. 2003. Osteoclast differentiation and activation. *Nature* **423:** 337–342.
34. Karsenty, G. & E.F. Wagner. 2002. Reaching a genetic and molecular understanding of skeletal development. *Dev. Cell* **2:** 389–406.
35. Teitelbaum, S.L. 2000. Bone resorption by osteoclasts. *Science* **289:** 1504–1508.
36. Tanaka, S. *et al.* 2005. Role of RANKL in physiological and pathological bone resorption and therapeutics targeting the RANKL-RANK signaling system. *Immunol. Rev.* **208:** 30–49.
37. Lacey, D.L. *et al.* 1998. Osteoprotegerin ligand is a cytokine that regulates osteoclast differentiation and activation. *Cell* **93:** 165–176.
38. Yasuda, H. *et al.* 1998. Osteoclast differentiation factor is a ligand for osteoprotegerin/osteoclastogenesis-inhibitory factor and is identical to TRANCE/RANKL. *Proc. Natl. Acad. Sci. USA* **95:** 3597–3602.
39. Asagiri, M. *et al.* 2005. Autoamplification of NFATc1 expression determines its essential role in bone homeostasis. *J. Exp. Med.* **202:** 1261–1269.
40. Takayanagi, H. *et al.* 2002. Induction and activation of the transcription factor NFATc1 (NFAT2) integrate RANKL signaling in terminal differentiation of osteoclasts. *Dev. Cell* **3:** 889–901.
41. Kitazawa, R. & S. Kitazawa. 2002. Vitamin D(3) augments osteoclastogenesis via vitamin D-responsive element of

mouse RANKL gene promoter. *Biochem. Biophys. Res. Commun.* **290:** 650–655.

42. Mizoguchi, F. *et al.* 2010. Osteoclast-specific Dicer gene deficiency suppresses osteoclastic bone resorption. *J. Cell Biochem.* **109:** 866–875.

43. Sugatani, T. & K.A. Hruska. 2009. Impaired micro-RNA pathways diminish osteoclast differentiation and function. *J. Biol. Chem.* **284:** 4667–4678.

44. Sugatani, T., J. Vacher & K.A. Hruska. 2011. A microRNA expression signature of osteoclastogenesis. *Blood* **117:** 3648–3657.

45. Mann, M. *et al.* 2010. miRNA-based mechanism for the commitment of multipotent progenitors to a single cellular fate. *Proc. Natl. Acad. Sci. USA* **107:** 15804–15809.

46. Yasui, T. *et al.* 2011. Regulation of RANKL-induced osteoclastogenesis by TGF-beta through molecular interaction between Smad3 and Traf6. *J. Bone Miner. Res.* **26:** 1447–1456.

Ann. N.Y. Acad. Sci. ISSN 0077-8923

ANNALS OF THE NEW YORK ACADEMY OF SCIENCES
Issue: *Skeletal Biology and Medicine II*

The osteoclast and its unique cytoskeleton

Steven L. Teitelbaum

Departments of Pathology and Immunology, and Medicine, Washington University School of Medicine, St. Louis, Missouri

Address for correspondence: Steven L. Teitelbaum, M.D., Washington University School of Medicine, Department of Pathology and Immunology, Department of Medicine, Campus Box 8118, 660 South Euclid Avenue, St. Louis, MO 63110. teitelbs@wustl.edu

The osteoclast cytoskeleton is a unique structure that polarizes the cell's resorptive machinery to the bone–cell interface where it creates an isolated resorptive microenvironment consisting of an actin ring surrounding a ruffled border. This polarization process occurs under the aegis of the $\alpha_v\beta_3$ integrin in collaboration with the M-CSF receptor, c-Fms. When occupied, $\alpha_v\beta_3$ activates a canonical signaling complex consisting of c-Src, Syk, Dap12, Slp76, Vav 3, and Rac that permits the cell to spread and form actin rings. Generation of the ruffled border, the cell's resorptive organelle, is an exocytic process wherein synaptotagmin VII mediates fusion of secretory lysosomes to the bone-apposed plasma membrane. Absence of any component of this signaling pathway compromises osteoclast cytoskeletal organization and abridges bone resorption.

Keywords: osteoclast; cytoskeleton; integrin; actin ring

The osteoclast, which is the sole established resorptive cell, degrades the skeletal matrix by forming a microenvironment between itself and the bone surface. It transports hydrogen and chloride ions into this isolated space, thereby creating an acidic milieu that mobilizes the skeleton's mineral phase. This process exposes the bone's organic matrix, consisting largely of type 1 collagen, which is subsequently degraded by cathepsin K.

Creation of the secluded microenvironment requires polarization of the osteoclast's resorptive machinery to the bone/cell interface and integrin-mediated physical intimacy with the extracellular matrix. Integrins are $\alpha\beta$ heterodimers, many of whose extracellular domains recognize matrix-residing proteins and whose intracellular components bind cytoskeleton-organizing and signaling molecules. While osteoclasts express a number of integrins, particularly those of the $\beta1$ family, the cell's principal bone-binding heterodimer is $\alpha_v\beta_3$ that recognizes the amino acid motif arginine-glycine-aspartic acid (RGD). This ligand sequence is present in a variety of skeleton-residing proteins including bone sialoprotein and osteopontin.

To determine the skeletal relevance of the integrin we deleted the $\beta3$ subunit gene in mice.[1] Os-teoclasts generated from macrophages derived from wild type and heterozygous animals are identical in that they stain for tartrate-resistant acid phosphatase (TRAP), are multinucleated and effectively spread in culture. While $\beta3$ homozygous-deficient cells also are multinucleated and express osteoclast differentiation markers they are incapable of spreading and appear "crenated" indicative of cytoskeletal dysfunction.

The osteoclast cytoskeleton is unique in that it forms a gasket-like structure known as the actin ring or sealing zone, which isolates the resorptive microenvironment from the general extracellular space. Actin rings are transient and present only when the cell is juxtaposed to bone. As the osteoclast detaches from the bone surface to access a new site of skeletal degradation, the circular structure disappears. Thus, organization of the osteoclast cytoskeleton is an essential component of the cell's capacity to resorb bone.

Cells lacking $\alpha_v\beta_3$ fail to form normal actin rings.[1,2] When occupied by its ligand, the matrix-recognizing heterodimer phosphorylates constitutively associated c-Src^{Y416} that activates the tyrosine kinase.[3] Activated cSrc, in turn, phosphorylates Syk whose SH2 domains bind the phosphotyrosines of

doi: 10.1111/j.1749-6632.2011.06283.x

the co-stimulatory ITAM protein, Dap12.[3,4] This $\alpha_v\beta_3$-associated complex recruits Slp76 that functions as an adaptor protein for Vav3.[5]

Vav3, which is relatively osteoclast-specific, is a guanine nucleotide exchange factor that transits small GTPases of the Rho family from their inactive GDP to their active GTP-bound form.[6] Cdc42 and Rac are two Rho family GTPases expressed in osteoclasts and we therefore asked if either participate in $\alpha_v\beta_3$-mediated organization of the cell's cytoskeleton. While cdc42 exerts a mild cytoskeletal effect, presumably through atypical PKCs, its predominant impact is on proliferation and apoptosis of osteoclast precursors and the mature cell, respectively.[7] On the other hand, Rac is an effector molecule of the $\alpha_v\beta_3$ integrin and its deletion results in severe compromise of the osteoclast cytoskeleton and, in consequence, impaired resorption.[8,9] In fact, deletion of Rac1 and Rac2, the isoforms present in the osteoclast, causes osteopetrosis more severe than that induced by absence of any component of the $\alpha_v\beta_3$ signaling complex, save c-Src.[9]

Having established an intracellular signaling pathway emanating from $\alpha_v\beta_3$ that organizes the osteoclast cytoskeleton and promotes bone resorption, we turned to the mechanism by which the integrin recognizes its ligand to activate the complex. In its resting, inactive state $\alpha_v\beta_3$ is characterized by a bent appearance of its external compartment and a close approximation of its intracellular domains. Upon activation the external domains of $\alpha_v\beta_3$ "straighten" and its intracellular tails separate, thereby inducing a high affinity ligand-binding state.

Integrins undergo either outside-in or inside-out activation. Outside-in activation typically follows ligand recognition by the inactive integrin, transmitting intracellular signals that induce its high affinity configuration. Inside-out activation, on the other hand, reflects signals derived from stimulated growth factor or cytokine receptors, which target the integrin's intracellular domains leading them to disrupt their salt bridge, separate and induce the extracellular region to undergo conformational change. To date, most studies of osteoclast integrins have focused on outside-in activation. In this circumstance, osteoclasts or their precursors are "parachuted" into wells containing an immobilized substrate recognized by the integrin's extracellular domain. The most significant mechanism by which integrins modify osteoclast function, however, is

likely inside-out stimulation by growth factors and cytokines thereby altering the cell's cytoskeleton and promoting bone-resorption.

Osteoclast differentiation occurs under the aegis of two essential cytokines, RANK ligand and M-CSF. Both, however, also promote actin remodeling of mature osteoclasts and enhance their capacity to resorb bone.[10,11] Given the importance of $\alpha_v\beta_3$ in organizing the cell's cytoskeleton, and evidence obtained in transformed cells,[12] we asked the integrin undergoes inside-out activation via M-CSF. We focused on this cytokine because it collaborates with $\alpha_v\beta_3$ in osteoclast cytoskeleton rearrangement as both stimulate the same c-Src-initiated signaling complex.[8,11] This observation raised the question as to whether the cytokine- and integrin-induced cytoskeletal effects are independent or cooperative wherein M-CSF activates $\alpha_v\beta_3$. To address the latter possibility, we asked if M-CSF-mediated Rac activation requires $\alpha_v\beta_3$.[8] We find that M-CSF induces Rac-GTP formation in wild type osteoclasts but not those lacking $\alpha_v\beta_3$. Because this experiment fortifies the hypothesis that the relationship between M-CSF and $\alpha_v\beta_3$ in organizing the osteoclast cytoskeleton reflects inside-out activation of the integrin by the cytokine, we asked if the cytokine alters the conformation of $\alpha_v\beta_3$ from its low to high affinity state. To this end, we used a monoclonal antibody that recognizes $\alpha_v\beta_3$ only in its high affinity conformation. Treatment of osteoclast lineage cells with M-CSF doubles the abundance of high affinity $\alpha_v\beta_3$.[2]

If, in fact, M-CSF induces high affinity conformation of the $\alpha_v\beta_3$ integrin, it is likely to do so by targeting the β_3 cytoplasmic domain. To determine if such is the case, we transduced β_3-null osteoclast precursors with either wild type β_3 subunit or that lacking its cytoplasmic tail. In contrast to the cytokine's capacity to enhance the abundance of high affinity, wild type $\alpha_v\beta_3$, absence of the β_3 cytoplasmic domain prohibits the cytokine from altering the integrin's conformational state.[2]

Having established that M-CSF induces high affinity conformation of $\alpha_v\beta_3$, we turned to the integrin-interacting molecule(s) induced by the cytokine. Our first candidate was talin that, in other circumstances, binds to integrins as a final common step in their activation. To explore this issue, we treated osteoclasts with M-CSF and immunoblotted β_3 subunit immunoprecipitates for talin and find it progressively associates with the integrin.

Because this observation indicates that talin may regulate osteoclast function, we crossed mice, in which the talin gene is floxed, to mice expressing Cre recombinase driven by the cathepsin K promoter. Establishing physiological relevance, mice in whom talin is conditionally deleted in osteoclasts have a marked increase in bone mass as detected radiographically, by μCT and histologically. The abundance of osteoclasts in these animals substantiates their skeletal abnormality is due to dysfunction of bone resorptive cells and not failed generation.

While osteoclasts are abundant in talin conditionally-deleted animals, they poorly adhere to the bone surface. Thus, M-CSF acting through its receptor, c-fms, promotes high affinity changes of the $\alpha_v\beta_3$ integrin by inducing talin association with the β3 cytoplasmic domain. The consequence of arresting this phenomenon is failed activation of the integrin-associated cytoskeleton-organizing complex leading to diminished bone resorption.

Bone resorption occurs under the aegis of a cytoskeletal structure of the osteoclast known as the ruffled border. This morphologically unique organelle, which consists of complex folds of plasma membrane juxtaposed to bone, appears only in osteoclasts adherent to the mineralized matrix and in a resorptive state. It is best visualized by electron microscopy that shows it surrounded by the actin ring or sealing zone. Thus, the resorptive machinery of the osteoclast consists of actin rings encompassing the ruffled border.

The ruffled border forms by polarization of cytoplasmic vesicles to the bone-apposed plasma membrane into which they insert, leading to its enhanced complexity. By this mechanism, the vesicles deliver an electrogenic H^+ATPase or proton pump and chloride channel, into the ruffled border thus providing the means to acidify the resorptive space, and cathepsin K, to degrade the organic matrix of bone. This process appears to mirror exocytosis that involves fusion of cytoplasmic vesicles to the plasma membrane, regulated, in part, by synaptotagmins. These calcium-sensing molecules are expressed in secretory vesicles and plasma membrane. Given the

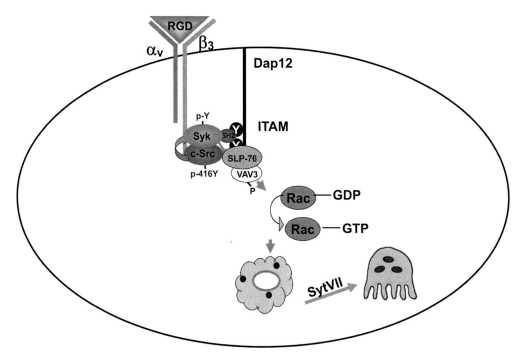

Figure 1. $\alpha_v\beta_3$ Integrin-mediated cytoskeletal organization. Upon occupancy by its RGD ligand, $\alpha_v\beta_3$ activates constitutively associated c-Src that in turn recruits and phosphorylates Syk. The SH2 domains of activated Syk interacts with Dap12 phosphotyrosines and recruits SLP-76 that functions as an adaptor for the guanine nucleotide exchange factor, VAV3. Syk-phosphorylated VAV3 transits Rac from its inactive GDP-bound state to its active GTP-bound state. GTP-Rac organizes the osteoclast cytoskeleton to generate actin rings via the Arp2/3 complex (not shown). When in contact with the mineralized matrix, the cell's resorptive organelle, the ruffled border, forms within the actin ring under the aegis of synaptotagmin VII.

importance of synaptotagmins, particularly the isoform synaptotagmin VII, in other cells undergoing
vesicle/membrane fusion, we postulated it partici-
pates in ruffled border formation. Supporting this
contention, synaptotagmin VII, transduced into osteoclasts as a GFP fusion protein, localizes within
the actin ring, thus in juxtaposition to the ruffled border.[13] Importantly, wild type osteoclasts localize cathepsin K to the resorptive microenvironment, whereas, the enzyme is diffusely distributed
throughout cytoplasm in those lacking synaptotagmin VII.

Evidence indicates that the ruffled border contains LAMP proteins, suggesting that the vesicles forming this complex membrane structure are
indeed secretory lysosomes.[14] Synaptotagmin VII
co-localizes with osteoclast lysosomes in the cytoplasm, consistent with the concept that it promotes fusion of these vesicles to the bone-apposed
plasma membrane, to generate the resorptive organelle.[13] Establishing such is the case, synaptotagmin VII$^{-/-}$ osteoclasts fail to form ruffled borders,
in vivo.

In face of the resorptive inefficiency of synaptotagmin VII deficient osteoclasts, one would expect increased bone mass in mice from whom the
cells were derived. Surprisingly, however, the bone
density of synaptotagmin VII-deficient mice is approximately 50% of their wild type counterparts, an
observation consistent with a greater suppression
of bone formation than resorption. Supporting this
hypothesis, bone nodule generation by synaptotagmin VII$^{-/-}$ osteoblasts is impaired.

Osteoblasts also accomplish their biological mis-
sion by secreting essential molecules, in this case,
collagen, osteocalcin, and transforming growth factor β, into the nascent bone matrix to promote its
production. Like that of osteoclasts, osteoblast secretion is regulated by synaptotagmin VII and in its
absence collagen release is abridged, resulting in intracellular procollagen accumulation. Establishing
physiological relevance, serum osteocalcin parallels
the decrease in bone mass, in synaptotagmin VII
knockout mice. Furthermore, histomorphometric
kinetic measurement establishes that mineral apposition and bone formation rates are diminished.
Thus, synaptotagmin VII regulates the secretory ac-

tivities of osteoclasts and osteoblasts. Absence of the
protein inhibits this function in both cell types but
more profoundly affects osteoblasts, eventuating in
osteoporosis (Fig. 1).

Conflicts of interest

The author declares no conflict of interest.

References

1. McHugh, K.P., K. Hodivala-Dilke, M.H. Zheng, *et al.* 2000. Mice lacking β3 integrins are osteosclerotic because of dysfunctional osteoclasts. *J. Clin. Invest.* **105:** 433–440.
2. Faccio, R., D.V. Novack, A. Zallone, *et al.* 2003. Dynamic changes in the osteoclast cytoskeleton in response to growth factors and cell attachment are controlled by β3 integrin. *J. Cell Biol.* **162:** 499–509.
3. Zou, W., H. Kitaura, J. Reeve, *et al.* 2007. Syk, c-Src, the αvβ3 integrin, and ITAM immunoreceptors, in concert, regulate osteoclastic bone resorption. *J. Cell Biol.* **176:** 877–888.
4. Zou, W., J.L. Reeve, H. Zhao, *et al.* 2009. Syk tyrosine 317 negatively regulates osteoclast function via the ubiquitin-protein isopeptide ligase activity of Cbl. *J. Biol. Chem.* **284:** 18833–18839.
5. Reeve, J.L., W. Zou, Y. Liu, *et al.* 2009. SLP-76 couples Syk to the osteoclast cytoskeleton. *J. Immunol.* **183:** 1804–1812.
6. Faccio, R., S.L. Teitelbaum, K. Fujikawa, *et al.* 2005. Vav3 regulates osteoclast function and bone mass. *Nat. Med.* **11:** 284–290.
7. Ito, Y., S.L. Teitelbaum, W. Zou, *et al.* 2010. Cdc42 regulates bone modeling and remodeling in mice by modulating RANKL/M-CSF signaling and osteoclast polarization. *J. Clin. Invest.* **120:** 1981–1993.
8. Faccio, R., S. Takeshita, A. Zallone, *et al.* 2003. c-Fms and the αvβ3 integrin collaborate during osteoclast differentiation. *J. Clin. Invest.* **111:** 749–758.
9. Croke, M., F.P. Ross, M. Korhonen, *et al.* 2011. Rac deletion in osteoclasts causes severe osteopetrosis. *J. Cell Sci.* (In press).
10. Lacey, D.L., E. Timms, H.L. Tan, *et al.* 1998. Osteoprotegerin ligand is a cytokine that regulates osteoclast differentiation and activation. *Cell* **93:** 165–176.
11. Zou, W., J.L. Reeve, Y. Liu, *et al.* 2008. DAP12 couples c-Fms activation to the osteoclast cytoskeleton by recruitment of Syk. *Mol. Cell* **31:** 422–431.
12. Faccio, R., M. Grano, S. Colucci, *et al.* 2002. Localization and possible role of two different αvβ3 integrin conformations in resting and resorbing osteoclasts. *J. Cell Sci.* **115:** 2919–2929.
13. Zhao, H., Y. Ito, J. Chappel, *et al.* 2008. Synaptotagmin VII regulates bone remodeling by modulating osteoclast and osteoblast secretion. *Dev. Cell* **14:** 914–925.
14. Baron, R., L. Neff, D. Louvard & P.J. Courtoy. 1985. Cell-mediated extracellular acidification and bone resorption: evidence for a low pH in resorbing lacunae and localization of a 100-kD lysosomal membrane protein at the osteoclast ruffled border. *J. Cell Biol.* **101:** 2210–2222.

Ann. N.Y. Acad. Sci. ISSN 0077-8923

ANNALS OF THE NEW YORK ACADEMY OF SCIENCES
Issue: *Skeletal Biology and Medicine II*

HDAC inhibitor trichostatin A suppresses osteoclastogenesis by upregulating the expression of C/EBP-β and MKP-1

Paul J. Williams, Kazi Nishu, and Md Mizanur Rahman

Department of Medicine, University of Texas Health Science Center at San Antonio, San Antonio, Texas

Address for correspondence: Md Mizanur Rahman, Department of Medicine, University of Texas Health Science Center at San Antonio, 7703 Floyd Curl Drive, San Antonio, TX 78229–3900. rahmanm@uthscsa.edu

Histone deacetylases (HDACs) remove the acetyl groups from the lysine residues of histone tails, leading to the formation of a condensed and transcriptionally silenced chromatin. HDAC inhibitors (HDACi) block this action and can result in hyperacetylation of histones, leading to a less compact and more transcriptionally active chromatin and thereby, gene expression. Previously, we have shown that HDACi inhibit osteoclast differentiation. However, which genes are transcriptionally activated following hyperacetylation of histones, and lead to the suppression of osteoclastogenesis, has yet to be elucidated. In this study, we show that an HDACi, trichostatin A (TSA), inhibits the receptor activator of the nuclear factor-κB (NF-κB) ligand (RANKL)–stimulated TNF-α production, NF-κB activation, and bone resorbing pit formation, and downregulates c-Fos and NFATc1 in RAW 264.7 cells. Interestingly, expression of antiosteoclastogenic factors CCAAT enhancer binding protein (C/EBP)-β and mitogen-activated protein kinase phosphatase (MKP)-1 was significantly upregulated in TSA-treated, RANKL-stimulated RAW 264.7 cells. These findings suggest that TSA upregulates the expression of C/EBP-β and MKP-1, which may downregulate pro-osteoclastogenic factors and signaling molecules, ultimately suppressing osteoclastogenesis.

Keywords: osteoclastogenesis; HDAC inhibitors; trichostatin A; C/EBP-β; MKP-1

Introduction

Bone remodeling depends on a delicate balance between bone formation and bone resorption. Tipping this balance in favor of bone resorption (osteoclasts) leads to pathologic bone diseases like osteoporosis and affects 10 million Americans over the age of 50 and results in 1.5 million fractures annually.[1] Osteoclasts are multinucleated cells that differentiate from myeloid precursors in response to the macrophage colony-stimulating factor and receptor activator of the nuclear factor-κB (NF-κB) ligand (RANKL). NFATc1 is a target of RANKL signaling. During osteoclastogenesis, NFATc1 expression is induced, and this transcription factor can be identified at the promoters of osteoclast-specific genes.[2] It has been shown that NFATc1 is the key to RANKL-regulated osteoclast differentiation.[3] RANKL also induces c-Fos by an as yet unknown mechanism. Indeed, the essential role of c-Fos pathway in osteoclastogenesis, as determined by gene targeting studies, is well documented.[4] c-Fos has also been found to be induced by RANKL–RANK signals.[4]

Histone acetylation, in concert with other histone modifications, has been described as a major epigenetic regulator for controlling cell fate. Histone acetyltransferases (HATs) transfer acetyl groups to core histones, resulting in local expansion of chromatin and increased accessibility of DNA binding proteins, leading to transcriptional activation.[5] Histone deacetylases (HDACs) are known to counteract the activity of HATs, thus functioning as transcriptional repressors. However, genome-wide mapping of HATs and HDACs reveals that both HATs and HDACs are associated in active genes with acetylated histones.[6] HDACs and HATs have histones and many nonhistone proteins as targets that have a

doi: 10.1111/j.1749-6632.2011.06286.x

Ann. N.Y. Acad. Sci. 1240 (2011) 18–25 © 2011 New York Academy of Sciences.

role in regulating gene expression.[7–9] There are two major roles for HDACs. One is their function in active genes, where high levels of HDACs act to remove the acetyl group added by high levels of HATs during the process of transcriptional initiation and elongation and a reset of the chromatin structure required for the next round of transcription. The other is their function in primed genes, where transient binding of HDACs removes the acetyl group that resulted from transient binding of HATs, maintains a low level of acetylation, and prevents Pol II binding, thereby maintaining promoters in an inactive state. In humans and mice, the 18 HDAC enzymes are grouped into four classes. Classical HDACs (class I, II, and IV) share sequence similarity and are dependent on Zn^{2+} for enzymatic activity, whereas the class III sirtuins act through a distinct NAD^+-dependent mechanism.[10] Class I HDACs (HDAC1, 2, 3, and 8) are generally localized to the nucleus and, with the exception of HDAC3, the absence of a nuclear export signal.[10] Class I HDACs have been most widely studied in their classical role as histone modifiers and transcriptional repressors. The class II enzymes have been subdivided into class IIa (HDAC4, 5, 7, and 9) and IIb (HDAC6 and 10) based on domain organization.[10] Class IIa HDACs possess N-terminal domains that interact with transcription factors. They also possess C-terminal nuclear export signals, which enable shuttling between the nucleus and cytoplasm. Nuclear export prevents class IIa HDACs from acting as transcriptional repressors, thus resulting in inducible gene expression. Class IIa HDACs primarily control gene expression by recruiting other proteins (corepressors or coactivators).[6,10] Class IIb HDACs (HDAC6 and HDAC10) are distinguished from the class IIa subfamily in possessing tandem deacetylase domains, although the second domain of HDAC10 is reported to be nonfunctional.[6] HDAC6 is unique among the classical HDAC family in that it is predominantly cytoplasmic, whereas HDAC10 is found in both the nucleus and cytoplasm.[6] HDAC11 is the sole class IV HDAC. HDACs, which target histones as well as nonhistone proteins as substrates have the potential to regulate gene expression. HDAC inhibitors (HDACi) are known to modulate the expression of genes by increasing histone acetylation and thereby, regulating chromatin structure and transcription.[11] Current evidence indicates that HDACi act not only to block the catalytic activity of the enzyme but

may also affect the protein–protein interaction of specific HDACs with various critical protein partners.[8] These target proteins are involved in many cell pathways including gene expression, cell proliferation, differentiation, cell migration and cell death, and have a role in angiogenesis and immune response. HDACi are emerging as a new class of potential therapeutic agents for the treatment of solid and hematologic malignancies.[12] HDACi have also recently been identified as potential antiarthritic agents. In animal models of rheumatoid arthritis, HDACi inhibited joint swelling, synovial inflammation, and subsequent bone and cartilage destruction.[13] HDACi include several structurally diverse natural products. Currently, there are several classes of HDACi, including butyrate, hydroxamic acid, benzamide, and cyclic peptides. A hydroxamic acid, trichostatin A (TSA), is a classical HDACi that blocks the activity of all isoforms with similar affinities, except class IIa HDACs, and was originally isolated as a fungistatic antibiotic from the *Streptomyces platensis* strain.[14] Anti-inflammatory properties of HDACi including TSA are well characterized.[12] Various HDACi are reported to enhance osteogenic differentiation in multiple cell types, including osteoblasts, bone marrow mesenchymal cells, and adipose-derived stromal cells.[15,16] Previously, we have shown that both sodium butyrate (NaB) and TSA are able to inhibit osteoclast differentiation in the monocyte macrophage cell line RAW 264.7 and mouse and rat primary bone marrow cell cultures.[17] HDACi may inhibit osteoclastogenesis by altering the structure of many proteins targets of HDACs, that is, histones and nonhistone proteins. However, because of hyperacetylation of histone and nonhistone proteins, transcriptional upregulation of those genes responsible for the suppression of osteoclast differentiation has yet to be elucidated. In this study, we determined that antiosteoclastogenic factors, CCAAT enhancer binding protein (C/EBP)-β and mitogen-activated protein kinase phosphatase (MKP)-1 are upregulated by TSA in RANKL-stimulated RAW 264.7 cells.

Materials and methods

All media components were purchased from GIBCO (Invitrogen Corp., Carlsbad, CA). TSA was purchased from Sigma (St. Louis, MO). Murine sRANKL was obtained from Pepro Tech, Inc. (Rocky Hill, NJ). The bicinchoninic acid kit for protein determination was from Pierce Chemical Co. (Rockford, IL). Rabbit polyclonal antibodies against NFATc1, p65 NF-κB; MKP-1, C/EBP-β and goat polyclonal antibody against actin, Lamin B and secondary antibodies were purchased from Santa Cruz Biotechnology (Santa Cruz, CA). Chicken polyclonal antibody against c-Fos was purchased from Abcam, Inc. (Cambridge, MA).

Cell cultures

RAW 264.7 cells were cultured in α-MEM medium (Sigma) with 10% fetal calf serum (FCS; Sigma). All media were supplemented with 2 mM glutamine, 100 IU/mL penicillin, and 100 mg/mL streptomycin (Sigma).

Pit formation assay

RAW 264.7 cells were suspended in α-MEM containing 10% FCS and plated at a concentration of 1×10^4 cells/well on an osteoclast activity assay substrate plate (OCT USA Inc., Chunan, Choongnam, Korea) in the presence or absence of 50 ng/mL RANKL and incubated for 24 hours. TSA (20 nM) was then added to the cultures. Half of the medium was replaced with a fresh media once every two days. After seven days of culture, the plates were washed in 6% sodium hypochlorite solution to remove the cells. The resorbed areas on the plates were captured with a digital camera attached to a microscope and analyzed by the Metaview Image Analysis system.[18] Resorbed pit area was expressed as % resorption area.

TNF-α production in RAW 264.7 cell culture

RAW 264.7 cells were suspended in α-MEM containing 10% FCS and plated at a concentration of 1×10^5 cells/well into a 24-well culture dish (Corning, NY) for 24 hours. They were then treated with/without TSA (10 and 20 nM) for an additional 24 hours before adding RANKL (50 ng/mL) to stimulate TNF-α secretion for 16 hours. Culture medium was collected and analyzed for TNF-α using the ELISA kit.[18]

Protein preparation and Western blot analysis

To determine the nuclear protein level of p65 NF-κB, RAW 264.7 cells were cultured in α-MEM with 10% FCS. After 24 hours of incubation, cells were treated with TSA (100 nM) for an additional 24 hours and then incubated with sRANKL (50 ng/mL) for several time points (0, 5, 10, 20, and 30 minutes). Cells were then collected and nuclear extracts were prepared,

as previously described.[17] For determining the levels of all other factors, RAW 264.7 cells were cultured in α-MEM with 10% FCS for 24 hours. sRANKL (50 ng/mL) and TSA (20 nM) were added and incubated for different time points as mentioned. Cells were collected and whole cell extracts and nuclear extracts were prepared as described earlier.[18] Protein concentrations of the nuclear extracts, cytosolic extracts, and whole cell extracts were determined using a bicinchoninic acid protein assay kit (Pierce, Rockford, IL). Fifty μg of whole cell extract and 15 μg of nuclear extract were subjected to Western blot analysis.

Statistical analysis

Results are expressed as mean ± SEM. Data was statistically analyzed using either Student's *t*-test or one-way ANOVA and $P < 0.05$ was considered statistically significant. Unpaired *t*-test and Newman–Keuls multiple comparison tests were used to test the differences between groups for significance using Graphpad prism for Windows (La Jolla, CA).

Results

Effect of TSA on bone resorption by RANKL

Earlier, we reported that TSA inhibited osteoclastogenesis.[17] Here we examined whether TSA has any effect on bone resorbing pit formation in RANKL-stimulated RAW 264.7 cells cultured in calcium phosphate-coated tissue culture plates. Cultures treated with TSA showed significantly reduced numbers and areas of resorption pits, compared with cultures treated with RANKL alone (Fig. 1A).

Effects of TSA on RANKL-stimulated TNF-α production

TNF-α is a strong pro-osteoclastogenic factor.[19,20] Therefore, we examined whether TSA can modulate RANKL-stimulated TNF-α production in RAW 264.7 cell cultures. Cultures treated with TSA exhibited dose-dependent reduction of TNF-α production compared with cultures treated with RANKL alone (Fig. 1B).

Effects of TSA on RANKL-stimulated NF-κB activation

Previously, we showed that TSA inhibits TNF-α-induced activation of NF-κB.[17] RANKL is the most pivotal factor mediating osteoclastogenesis. Therefore, we examined the effect of TSA on RANKL-stimulated activation of NF-κB by determining the

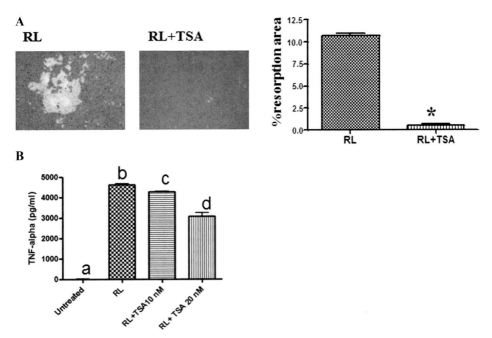

Figure 1. TSA inhibited RANKL (RL)-stimulated bone resorption and TNF-α production of RAW 264.7 cells. (A) Demonstration of pit formation on an osteoclast activity assay substrate plate using RAW 264.7 cells stimulated with 50 ng/mL RANKL with or without 20 nM TSA. TSA was added after 24 hours of incubation of RAW 264.7 cells with RANKL. After seven days of culture, plates were analyzed for resorption pit formation. Bar diagram shows the percentage resorption area per well measured by the image analyzer. Values shown are means ± SEM of three independent experiments with triplicate cultures. Statistical differences were evaluated between the groups with a Student's *t*-test. *Significantly different at $P < 0.001$. (B) Dose-dependent inhibition of RANKL (RL)-induced TNF-α production of RAW 264.7 cells by TSA. RAW 264.7 cells were cultured for 24 hours. Then they were treated with/without TSA for another 24 hours before adding RANKL (50 ng/mL) to stimulate TNF-α secretion for 16 hours. Medium was collected and analyzed for TNF-α using an ELISA kit. Each value represents the mean ± SEM of three independent quadruplicate cultures. Statistical differences were evaluated among all groups with one-way ANOVA. Value with different superscripts are significantly different at $P < 0.05$ by Newman–Keuls' one-way ANOVA with multiple comparison test.

nuclear levels of the p65 NF-κB subunit. Interestingly, dramatic reduction of nuclear p65 NF-κB level was found in TSA-treated RAW 264.7 cells (Fig. 2A). These data support our previous findings that TSA inhibits RANKL-stimulated NF-κB–dependent transactivation in RAW 264 cells.[17]

Effects of TSA on RANKL-stimulated expression of NFATc1 and c-Fos

NFATc1 and c-Fos are considered to be the most important osteoclast-specific transcription factors and the master regulators of osteoclastogenesis.[2,21] We examined whether TSA modulates these transcription factors in RANKL-stimulated RAW 264.7 cells. RAW 264.7 cells treated with TSA significantly decreased the RANKL-stimulated upregulation of NFATc1 and c-Fos transcription factors (Fig. 2B). However, RAW 264.7 cell treated with only TSA did

not have any effect on the expression of these transcription factors (data not shown).

Effects of TSA on RANKL-stimulated C/EBP-β expression

Very recently, C/EBP-β has been reported to be involved in controlling osteoclastogenesis.[22] Therefore, we determined if TSA upregulated the expression of C/EBP-β in RAW 264.7 cells treated with RANKL. We first determined the levels of C/EBP-β during the course of osteoclast differentiation in RAW 264.7 cells in the presence of RANKL. We found that expression of C/EBP-β decreased with time during the course of RANKL-induced osteoclast differentiation (Fig. 3A). We determined the levels of C/EBP-β at zero hour, six hours, and five days time points. At six hours' incubation, osteoclast formation does not occur in RAW 264.7 cells in the presence of RANKL, whereas at five days a

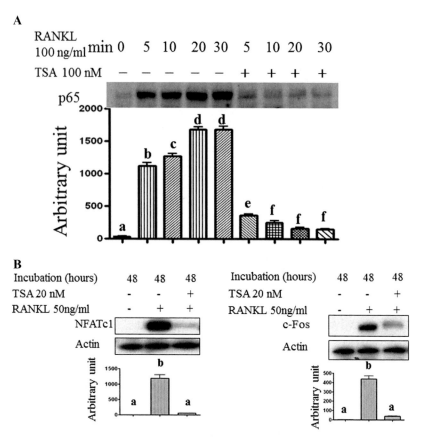

Figure 2. TSA reduced the expression of p65 NF-κB and NFATc1 and c-Fos in RAW 264.7 cells stimulated with RANKL. (A) RAW 264.7 cells were treated without or with TSA (100 nM) for 24 hours, followed by treatment with RANKL for 5, 10, 20, and 30 minutes. Nuclear extracts were prepared and 15 μg of proteins were analyzed by Western blotting using p65 NF-κB antibody. (B) RAW 264.7 cells were cultured for 24 hours and then RANKL (50 ng/mL) and TSA were added and cultured for another 48 hours. Whole cell lysates were prepared, and 50 μg of proteins were analyzed by Western blotting using antibodies to nuclear factor of activated T cells (NFATc1) and c-Fos. Lower panel shows the loading control (actin) for whole cell lysates. The intensity of the bands was determined by densitometry. Values shown are means ± SEM of two independent experiments. Statistical differences were evaluated among all groups with one-way ANOVA. Value with different superscripts are significantly different at $P < 0.05$ by Newman–Keuls' one-way ANOVA with a multiple comparison test.

number of osteoclasts formed when RAW 264.7 cells were treated with RANKL. Interestingly, we did not find any downregulation of C/EBP-β in the six hours RANKL-treated culture, whereas dramatic inhibition of C/EBP-β expression was noted in the five days RANKL-treated RAW 264.7 cells. However, TSA inhibited this downregulation of C/EBP-β expression in RAW 264.7 cells treated with RANKL (Fig. 3B).

Effects of TSA on RANKL-stimulated MKP-1 expression

MKP-1 inactivates mitogen-activated protein kinases (MAPKs), which are very important signaling molecules for osteoclastogenesis.[23,24] Therefore,

we determined if TSA-induced hyperacetylation of histone can upregulate the expression of MKP-1 in RAW 264.7 cells. Interestingly, TSA inhibits this downregulation of MKP-1 expression in RAW 264.7 cells treated with RANKL (Fig. 4).

Discussion

Here we show that two antiosteoclastogenic factors, C/EBP-β and MKP-1, were upregulated in RANKL-stimulated RAW 264.7 cells in the presence of TSA. Thus, HDAC inhibition may inhibit osteoclastogenesis by upregulating the expression of C/EBP-β and MKP-1. In this study, we showed that TSA inhibited bone resorption in a RAW 264.7 cell

A

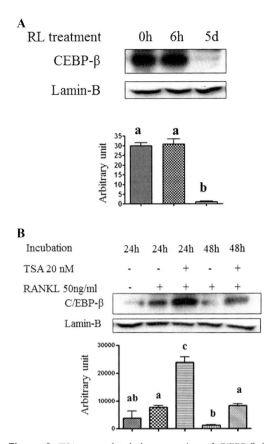

Figure 3. TSA upregulated the expression of C/EBP-β in RANKL (RL)-stimulated RAW 264.7 cells. (A) RAW 264.7 cells were cultured in the presence of RANKL (50 ng/mL) for five days. Nuclear extracts were prepared at indicated time points. (B) RAW 264.7 cells were cultured for 24 hours and then RANKL (50 ng/mL) and TSA were added and cultured for 24 and 48 hours. Nuclear extracts were prepared, and equal amounts of protein (15 μg) were analyzed by Western blotting using antibodies to C/EBP-β. Lower panel shows the loading control (Lamin B) for nuclear extracts. The intensity of the bands was determined by densitometry. Values shown are means ± SEM of two independent experiments. Statistical differences were evaluated among all groups with a one-way ANOVA. Values with different superscripts are significantly different at $P < 0.05$ by Newman–Keuls' one-way ANOVA with a multiple comparison test.

culture. The reduction of bone resorbing pit formation might be the direct result of the reduced number of osteoclast formations. Previously, we showed that TSA inhibits TNF-α–induced activation of NF-κB.[17] Because RANKL is a member of the TNF-α super family and the most important osteoclastogenic factor, here we showed that TSA could also inhibit RANKL-induced activation of NF-κB. These observations demonstrate that TSA

can inhibit both RANKL-dependent and RANKL-independent activation of NF-κB. Others have reported that TSA decreases the mRNA and protein levels of the proinflammatory cytokines, such as TNF-α, IL-6, and IL-1β, and in contrast, increases the level of the anti-inflammatory cytokine IL-10 in models of autoimmune and inflammatory diseases.[25] Our observations, together with these data, demonstrate that TSA is fundamentally anti-inflammatory in nature.

Previously, we have shown that TSA inhibits osteoclastogenesis by inhibiting the activation of NF-κB and p38 MAPK signaling. In this study, we found that TSA also reduced RANKL-stimulated upregulation of both c-Fos and NFATc1expression, the most pivotal osteoclastogenic transcriptional factors, in RAW 264.7 cells.[21,26] In general, HDAC inhibition leads to transcriptional activation. Therefore, the downregulation of c-Fos and NFATc1 may not be a direct effect of TSA, but rather an indirect effect of TSA, through the upregulation of factors, that negatively regulate the transcriptional activation of c-Fos and NFATc1.

In searching for the antiosteoclastogenic factors upregulated in TSA-treated RAW 264.7 cells, we first examined the expression of C/EBP-β. Recently, the action of C/EBP-β in mesenchymal cell differentiation, including osteoblasts has been addressed in several studies.[26–28] The general function of C/EBP-β in bone homeostasis, including its potential role in osteoclasts, has also been reported. C/EBP-β deficiency enhances bone resorption and exhibits enhanced osteoclast markers.[27] Smink *et al.* reported that C/EBP-β controls osteoclastogenesis through mTOR-mediated isoforms; liver-enriched activating protein (LAP), which acts as a transcriptional activator; liver-enriched inhibitory protein (LIP), which acts as a dominant negative inhibitor of LAP); and MafB, a negative regulator of osteoclastogenesis.[22] C/EBP-β mutant mouse strains exhibit increased bone resorption and attenuate expression of MafB.[22] These observations by others demonstrate that the effect of C/EBP-β on osteoclastogenesis may depend on LAP and LIP isoform switching. On the contrary, it has been reported that C/EBP-β and NF-κB cooperatively induce inflammatory gene products.[29,30] This effect might also be isoform specific. In our study, we observed that the expression of C/EBP-β decreases with time during the course of RANKL-induced osteoclast differentiation,

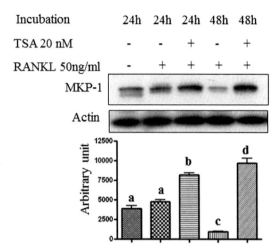

Figure 4. TSA upregulated the expression of MKP-1 in RANKL (RL)-stimulated RAW 264.7 cells. RAW 264.7 cells were cultured for 24 hours and then RANKL (50 ng/mL) and TSA were added and cultured for 24 and 48 hours. Whole cell lysates were prepared, and equal amounts of protein (50 μg) were analyzed by Western blotting using antibodies to MKP-1. The intensity of the bands was determined by densitometry. Values shown are means $\pm$ SEM of two independent experiments. Statistical differences were evaluated among all groups with one-way ANOVA. Values with different superscripts are significantly different at $P < 0.05$ by Newman–Keuls' one-way ANOVA with a multiple comparison test.

indicating that diminished expression of C/EBP-β may be a prerequisite for osteoclastogenesis. Sustained expression of C/EBP-β in RANKL-stimulated RAW 264.7 cells, in the presence of TSA, may be one of the mechanisms how TSA suppresses osteoclastogenesis. However, further studies are necessary to determine the effect of TSA on C/EBP-β isoform switching.

We then determined the effect of TSA on the expression of another antiosteoclastogenic factor MKP-1 and found upregulation of MKP-1 expression by TSA in RANKL-treated RAW 264.7 cells. MKP-1 dephosphorylates MAPKs.[31] The promoter of genes for MAPK phosphatase is expressed in macrophages and is activated by TSA.[32] Activation of MAPKs such as p38 and c-Jun N-terminal kinase (JNK) are associated with differentiation and activation of osteoclasts.[23,24] MKP-1 acts as a negative regulator of p38 and JNK. Macrophages from Mkp-1 knockout mice exhibited prolonged activation of p38 and JNK.[31] $Mkp-1^{-/-}$ macrophages produced more proinflammatory cytokines, such as TNF-α and IL-6, which are also known to be asso-

ciated with osteoclastogenesis.[31,33,34] Inhibition of proinflammatory cytokine biosynthesis by MKP-1 in LPS-stimulated macrophages has also been reported.[35,36] Furthermore, HDACi reduced inflammation and mortality in WT mice treated with LPS, but failed to protect MKP-1 knockout mice.[37] MKP-1 may limit the inflammatory reaction by inactivating JNK and p38, thus preventing osteoclast differentiation and activation caused by exaggerated inflammatory responses. Further, there is evidence that C/EBP-β plays an important role on transcriptional activation of MKP-1.[38]

In summary, TSA induces the expression of C/EBP-β and MKP-1, which may independently or synergistically suppress the osteoclast differentiation. The data in this study was obtained solely in the RAW cell line system. Further experiments using siRNA and chromatin immunoprecipitation technology using primary bone marrow cells and transgenic and knockout mice are necessary to confirm the activation of C/EBP-β and MKP-1 by HDACi. A PCR array should be done to find out other anti-inflammatory genes upregulated by HDACi and possible correlation with osteoclast inhibition. Further studies are also necessary to determine how HDACi modulates the nonhistone substrates and the relationship of these nonhistone substrate modulations in osteoclastogenesis. HDACi are promising compounds to treat osteoporosis and other inflammatory bone diseases.

Acknowledgments

This work was supported by NIH-K01-AG034233–02. We acknowledge Dr. Anthony Valente for his kind review of our manuscript for scientific and grammatical corrections.

Conflicts of interest

The authors declare no conflict of interest.

References

1. Aliprantis, A.O. *et al.* 2008. NFATc1 in mice represses osteoprotegerin during osteoclastogenesis and dissociates systemic osteopenia from inflammation in cherubism. *J. Clin. Invest.* **118:** 3775–3789.
2. Takayanagi, H. *et al.* 2002. Induction and activation of the transcription factor NFATc1 (NFAT2) integrate RANKL signaling in terminal differentiation of osteoclasts. *Dev. Cell.* **3:** 889–901.
3. Ikeda, F. *et al.* 2004. Critical roles of c-Jun signaling in regulation of NFAT family and RANKL-regulated osteoclast differentiation. *J. Clin. Invest.* **114:** 475–484.

4. Grigoriadis, A.E. *et al.* 1994. c-Fos: a key regulator of osteoclast-macrophage lineage determination and bone remodeling. *Science* **266:** 443–448.

5. Mottet, D. & V. Castronovo. 2008. Histone deacetylases: target enzymes for cancer therapy. *Clin. Exp. Metastasis* **25:** 183–189.

6. Wang, Z. *et al.* 2009. Genome-wide mapping of HATs and HDACs reveals distinct functions in active and inactive genes. *Cell* **138:** 1019–1031.

7. Shahbazian, M.D. & M. Grunstein. 2007. Functions of site-specific histone acetylation and deacetylation. *Annu. Rev. Biochem.* **76:** 75–100.

8. Marks, P.A. 2010. Histone deacetylase inhibitors: a chemical genetics approach to understanding cellular functions. *Biochim. Biophys. Acta* **1799:** 717–725.

9. Spange, S. *et al.* 2009. Acetylation of non-histone proteins modulates cellular signalling at multiple levels. *Int. J. Biochem. Cell Biol.* **41:** 185–198.

10. Shakespear, M.R. *et al.* 2011. Histone deacetylases as regulators of inflammation and immunity. *Trends Immunol.* **32:** 335–343.

11. Richon, V.M. & J.P. O'Brien. 2002. Histone deacetylase inhibitors: a new class of potential therapeutic agents for cancer treatment. *Clin. Cancer Res.* **8:** 662–664.

12. Dinarello, C.A. 2010. Anti-inflammatory agents: present and future. *Cell* **140:** 935–950.

13. Chung, Y.L. *et al.* 2003. A therapeutic strategy uses histone deacetylase inhibitors to modulate the expression of genes involved in the pathogenesis of rheumatoid arthritis. *Mol. Ther.* **8:** 707–717.

14. Yoshida, M. *et al.* 1990. Potent and specific inhibition of mammalian histone deacetylase both in vivo and in vitro by trichostatin A. *J. Biol. Chem.* **265:** 17174–17179.

15. Cho, H.H. *et al.* 2005. Induction of osteogenic differentiation of human mesenchymal stem cells by histone deacetylase inhibitors. *J. Cell Biochem.* **96:** 533–542.

16. Xu, Y. *et al.* 2009. Inhibition of histone deacetylase activity in reduced oxygen environment enhances the osteogenesis of adipose-derived stromal cells. *Tissue Eng. Part A* **15:** 3697–3707.

17. Rahman, M.M. *et al.* 2003. Two histone deacetylase inhibitors, trichostatin A and sodium butyrate, suppress differentiation into osteoclasts but not into macrophages. *Blood* **101:** 3451–3459.

18. Rahman, M.M., A. Bhattacharya & G. Fernandes. 2006. Conjugated linoleic acid inhibits osteoclast differentiation of RAW264.7 cells by modulating RANKL signaling. *J. Lipid Res.* **47:** 1739–1748.

19. Kobayashi, K. *et al.* 2000. Tumor necrosis factor alpha stimulates osteoclast differentiation by a mechanism independent of the ODF/RANKL-RANK interaction. *J. Exp. Med.* **191:** 275–286.

20. Lam, J. *et al.* 2000. TNF-alpha induces osteoclastogenesis by direct stimulation of macrophages exposed to permissive levels of RANK ligand. *J. Clin. Invest.* **106:** 1481–1488.

21. Wang, Z.Q. *et al.* 1992. Bone and haematopoietic defects in mice lacking c-fos. *Nature* **360:** 741–745.

22. Smink, J.J. *et al.* 2009. Transcription factor C/EBPbeta isoform ratio regulates osteoclastogenesis through MafB. *EMBO J.* **28:** 1769–1781.

23. Srivastava, S. *et al.* 2001. Estrogen decreases osteoclast formation by down-regulating receptor activator of NF-kappa B ligand (RANKL)-induced JNK activation. *J. Biol. Chem.* **276:** 8836–8840.

24. Matsumoto, M. *et al.* 2000. Involvement of p38 mitogen-activated protein kinase signaling pathway in osteoclastogenesis mediated by receptor activator of NF-kappa B ligand (RANKL). *J. Biol. Chem.* **275:** 31155–31161.

25. Han, S.B. & J.K. Lee. 2009. Anti-inflammatory effect of Trichostatin-A on murine bone marrow-derived macrophages. *Arch. Pharm. Res.* **32:** 613–624.

26. Zanotti, S. *et al.* 2009. Misexpression of CCAAT/enhancer binding protein beta causes osteopenia. *J. Endocrinol.* **201:** 263–274.

27. Motyl, K.J. *et al.* 2011. CCAAT/enhancer binding protein {beta}-deficiency enhances type 1 diabetic bone phenotype by increasing marrow adiposity and bone resorption. *Am. J. Physiol. Regul. Integr. Comp. Physiol.* **300:** R1250–R1260.

28. Tominaga, H. *et al.* 2008. CCAAT/enhancer-binding protein beta promotes osteoblast differentiation by enhancing Runx2 activity with ATF4. *Mol. Biol. Cell* **19:** 5373–5386.

29. Ray, A., M. Hannink & B.K. Ray. 1995. Concerted participation of NF-kappa B and C/EBP heteromer in lipopolysaccharide induction of serum amyloid A gene expression in liver. *J. Biol. Chem.* **270:** 7365–7374.

30. Liu, S. *et al.* 2011. Lingual antimicrobial peptide and IL-8 expression are oppositely regulated by the antagonistic effects of NF-kappaB p65 and C/EBPbeta in mammary epithelial cells. *Mol. Immunol.* **48:** 895–908.

31. Wang, X. *et al.* 2007. Knockout of Mkp-1 enhances the host inflammatory responses to gram-positive bacteria. *J. Immunol.* **178:** 5312–5320.

32. Musikacharoen, T., Y. Yoshikai & T. Matsuguchi. 2003. Histone acetylation and activation of cAMP-response element-binding protein regulate transcriptional activation of MKP-M in lipopolysaccharide-stimulated macrophages. *J. Biol. Chem.* **278:** 9167–9175.

33. Rahman, M. *et al.* 2009. Conjugated linoleic acid (CLA) prevents age-associated skeletal muscle loss. *Biochem. Biophys. Res. Commun.* **383:** 513–518.

34. Rahman, M.M. *et al.* 2009. Endogenous n-3 fatty acids protect ovariectomy induced bone loss by attenuating osteoclastogenesis. *J. Cell. Mol. Med.* **13:** 1833–1844.

35. Chen, P. *et al.* 2002. Restraint of proinflammatory cytokine biosynthesis by mitogen-activated protein kinase phosphatase-1 in lipopolysaccharide-stimulated macrophages. *J. Immunol.* **169:** 6408–6416.

36. Zhao, Q. *et al.* 2005. The role of mitogen-activated protein kinase phosphatase-1 in the response of alveolar macrophages to lipopolysaccharide: attenuation of proinflammatory cytokine biosynthesis via feedback control of p38. *J. Biol. Chem.* **280:** 8101–8108.

37. Cao, W. *et al.* 2008. Acetylation of mitogen-activated protein kinase phosphatase-1 inhibits Toll-like receptor signaling. *J. Exp. Med.* **205:** 1491–1503.

38. Johansson-Haque, K., E. Palanichamy & S. Okret. 2008. Stimulation of MAPK-phosphatase 1 gene expression by glucocorticoids occurs through a tethering mechanism involving C/EBP. *J. Mol. Endocrinol.* **41:** 239–249.

Ann. N.Y. Acad. Sci. ISSN 0077-8923

Skeletal receptors for steroid-family regulating glycoprotein hormones

A multilevel, integrated physiological control system

Harry C. Blair,[1] Lisa J. Robinson,[1] Li Sun,[2] Carlos Isales,[3] Terry F. Davies,[2] and Mone Zaidi[2]

[1]Pittsburgh VA Medical Center, University of Pittsburgh, Pittsburgh, Pennsylvania. [2]The Mount Sinai Bone Program, Mount Sinai School of Medicine, New York, New York. [3]Department of Medicine, Medical College of Georgia, Augusta, Georgia

Address for correspondence: Harry C. Blair, Pittsburgh VA Medical Center, University of Pittsburgh, Pittsburgh, PA 15261. hcblair@imap.pitt.edu

Pituitary glycoprotein hormone receptors, including ACTH-R, TSH-R, and FSH-R, occur in bone. Their skeletal expression reflects that central endocrine control is evolutionarily recent. ACTH receptors, in osteoblasts or the adrenal cortex, drive VEGF synthesis. VEGF is essential to maintain vasculature. In bone, ACTH suppression by glucocorticoids can cause osteonecrosis. TSH receptors occur on osteoblasts and osteoclasts, in both cases reducing activity. Thus, TSH directly reduces skeletal turnover, consistent with evolutionary adaptation to stress. FSH receptors accelerate bone resorption, whereas estrogen promotes bone formation, the forces usually balancing. With ovarian failure, low estrogen with high FSH causes rapid bone loss. The skeletal FSH effect in the menopause seems paradoxical, but it is a logical adaptation in lactation, where prolonged FSH elevation also occurs. In addition to receptors, there is some synthesis of pituitary glycoproteins at distributed sites; this is not well studied, but it may further modify the paradigm of central endocrine regulation.

Keywords: adrenocorticotropin; follitropin; G protein–coupled receptor; thyrotropin

The seven-transmembrane domain G protein–coupled receptors (GPCRs), including the pituitary hormone receptors, have extracellular loops that bind their ligands. When ligand binding occurs, the membrane-spanning regions are rearranged, activating associated GTP-binding regulatory proteins.[1] The GPCRs are a subclass of the largest group of receptors, the retinylidine-type receptor proteins, or rhodopsins, found essentially in all species. The advanced, or type 2 receptors,[2] which include the GPCRs, occur only in eukaryotes where they have proliferated and specialized. They include vertebrate photoreceptors and many chemoreceptor families, which are good examples of gene duplication and functional divergence. Many of the GPCRs transduce only local signals, such as in olfactory and taste organs, to activate neural signals.

Here, we will discuss three less specialized and more broadly distributed receptors: the TSH-R, named for binding the thyroid-stimulating hormone; the follicle-stimulating hormone receptor, FSH-R, which regulates maturation of germ cells and related processes; and the adrenocorticotropic hormone receptor, ACTH-R (also called MC2R). We selected these because they occur in the skeleton (Fig. 1) and because they have important, fairly well-characterized actions in the skeleton. These are good examples of a new paradigm, that the pituitary–endocrine axis is integrated with evolutionarily older, direct metabolic regulation mediated by the same receptors. There are many other skeletal receptors of the type 2-rhodopsin class, including parathyroid hormone, oxytocin, and frizzled receptors, that to varying extents signal in response to local and circulating hormones and are the subject of active investigation.[3–5] These will not be discussed extensively, but it should also be noted that direct central regulation of nonendocrine organs is not unique to regulation of bone.

doi: 10.1111/j.1749-6632.2011.06287.x

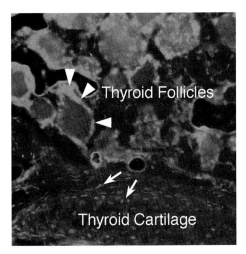

Figure 1. Replacement of an exon of the TSH-R with green fluorescent protein[21] reveals, in fluorescence microscopy of cross-sections of the thyroid of the heterozygote, that TSH-R expressing cells occur in the perichondrium of the thyroid cartilage and in differentiating chondrocytes (arrows, bottom), as well as through the expected strong green fluorescent protein expression in the thyroid follicular cells (arrowheads). Unlike TSH-R, the green fluorescent protein is a soluble protein, distributed throughout the cytoplasm. Control sections of homozygous thyroid and thyroid cartilage were negative (not shown).

Modification of an older, distributed metabolic regulatory system

ACTH is the clearest example of a pituitary hormone that is a specialized product that is part of an ancient and widely distributed GPCR system. ACTH-related signals and receptors participate in local cell differentiation in several contexts, but pituitary–adrenal signaling usually overshadows the distributed functions. There are five melanocortin receptors, including the ACTH receptor (MC2R).[6] They regulate cellular functions including pigment production in the skin, apatite, and sexual function. They are all controlled by ligands processed from a single large prohormone, pro-opiomelanocorticin (POMC). Hormone production occurs by tissue-specific regulated proteolysis,[7] with pH in part regulating cleavage.[8] The predominant product of the anterior pituitary is ACTH, although β-lipotropin, another POMC product, is also important. In other sites, POMC is also expressed but different processed products are made, including three melanotropins and β-endorphin. An excellent and recently discovered illustration that decentralized control is adapted for central control is that corticotropin-releasing factor (CRF), which in the adult stimulates ACTH production in the pituitary via a local vascular connection, occurs in general circulation in the fetus, and in that context, CRF stimulates cortisol synthesis directly.[9] This fetal response suggests that centralized ACTH production as the second messenger for CRF has not yet completely supplanted an ancestral regulatory system.

TSH and FSH are two of a group of hormones, along with chorionic gonadotropin (hCG) and leuteinizing hormone (LH), that have heterodimeric protein ligands that share a common α-chain. Their specificity depends on different, but related, β-chains, those of TSH, FSH, LH, and hCG being products of gene duplication and divergence, and retaining homologies, with the TSHβ chain having the greatest divergence. These hormones are particularly interesting in that, in simpler phyla, they have distributed functions. In coelenterates, which have a primitive nervous system but no endocrine glands, a TSH-R family gene is readily identifiable, widely expressed and shows the intron–exon structure found in mammals.[10] In the lower vertebrates, the TSH-R family has been carefully studied, with interesting findings including that, in bony fish, TSH-R is abundant in the thyroid and in ovaries, and is detectable in several other tissues including heart, muscle, and brain.[11] In fish, gonadal expression of LH-R and FSH-R is established, and all higher orders retain this. But curiously, multiple differently processed forms of the FSH-R occur in fish,[12] which may reflect receptors with different functions, as discussed below. Further, in the fish the FSH-R binds both FSH and LH, whereas the LH-R recognizes only LH.[13] In general, several workers found strong FSH-R expression only in gonads, but low level expression of FSH-R occurs in spleen of fish.[14] This, as will be discussed, is remarkably similar to findings in human cells.

ACTH regulates vascular survival and growth in bone

Glucocorticoids, under natural regulation, mainly by ACTH, are important coregulators of many processes including vascular tone, central metabolism, and immune response. At higher, pharmacological, levels, glucocorticoids are important tools in the management of diseases where control of the inflammatory response is important. These include ulcerative colitis, kidney diseases, and rheumatoid disorders. However, at concentrations greater than those produced by the adrenals under the regulation of ACTH, a variety of metabolic and medical complications occur. These include diabetes, osteoporosis, and osteonecrosis, a painful debilitating condition that affects metabolically active bone, most commonly the femoral head.[15] Glucocorticoids also cause a global skeletal loss which reflects a relative increase in bone resorption relative to bone formation. Osteoporosis is caused by net bone loss, whereas osteonecrosis reflects regional bone death with subsequent collapse. Osteonecrosis almost invariably requires surgical treatment. It leads to ∼50,000 hip replacements annually in the United States.[16] Various mechanisms, including vascular emboli and osteocyte apoptosis, have been promoted as causal determinants of glucocorticoid-induced osteonecrosis,[15] and a variety of empirical prevention strategies have been proposed, mainly aimed at hypotheses relating to abnormalities in fat metabolism and vascular integrity that occur with high glucocorticoid levels. Tests of these hypotheses have shown inconsistent results, reflecting that the underlying mechanisms probably are only partially, at most, related to those secondary changes with glucocorticoid treatment.

Detailed examination of the biochemical connection between glucocorticoid therapy and osteonecrosis recently was discovered to reflect, at least in large part, a direct bone response to ACTH. Key findings included the identification that necrosis of the bone forming unit (osteon) occurs before macroscopic vascular changes,[17] and the discovery by Isales and colleagues,[18] that bone forming units express the ACTH receptor strongly. ACTH induces vascular endothelial growth factor (VEGF) production in maintenance of the adrenal cortex and, we discovered, also in bone,[19] at least during bone synthesis. This discovery reconciles several observations that previously were difficult to connect with a logical mechanism. Specifically, we showed that ACTH protects against osteonecrosis of the femoral head in rabbits, which is induced by depot methylprednisolone acetate (Fig. 2).[19] This effect is directly linked to ACTH-induced production of VEGF by osteoblasts.[19] VEGF is a key factor maintaining the fine vascular network that supports bone synthesis. An independent report with consistent findings was recently published.[20] Overall, this work suggests that VEGF suppression secondary to glucocorticoid-induced ACTH suppression is a key factor driving bone damage with long-term glucocorticoid therapy. Production of ACTH outside the pituitary may in some circumstances be important; this is discussed separately below.

TSH signals regulate bone turnover activity

TSH response in bone is the best established of the pituitary GPCH receptors. This is owed, in part, to the very interesting mouse with part of the TSH-R replaced by green fluorescent protein (GFP; Fig. 1).[21] When the thyroid of the heterozygotic animal for the GFP transgene replacing TSH-R function but with one normal TSH-R gene, it was seen that the expected high expression of TSH-R occurred not only in thyroid follicles, but in cells of the perichondrium and thyroid cartilage. Although considerable analysis remains to be done, TSH-R seem to be expressed in most skeletal cells, and activity of osteoblasts and osteoclasts both are increased by TSH.[22] The effect on bone degradation is consistent with reduced cell differentiation, and a large part of the effect may be because of TSH inhibiting the production TNF-α in bone. Details of this mechanism, although it is strongly supported by knockout mouse experiments,[23] are not fully understood. Indeed, some types of marrow cells including macrophage-depleted (CD11b negative) marrow cells may respond to TSH by decreased TNF-α expression.[24] Production of TSH by immune cells may also regulate bone response, in addition to pituitary regulation; possible roles of extra-pituitary ligand production are discussed separately later.

FSH drives bone resorption

The finding of FSH receptors in bone cells, principally in cells of the monocytic lineage, remains controversial, and indeed the FSH-R is a low-level product that probably mainly is an unusual receptor isoform.[25] The evidence from knockout mouse work and *in vitro* differentiation indicates that FSH-R in bone, in keeping with an atypical isoform, signals via $G_{i2\alpha}$ and activates Erk, Akt, and NF-κB pathways rather than the usual cAMP ($G_{\alpha s}$) pathway of the FSH-R in ovary.[26] Alternatively spliced FSH-R isoforms and low-level expression in spleen had independently been described in osteichthyes, as discussed in the evolutionary perspective above.[12,14] Further findings include that FSH drives TNF-α production in bone,[27] suggesting that a TNF-α mechanism operates in opposition to the TSH effect,[23] and thus that a common TNF-α pathway may

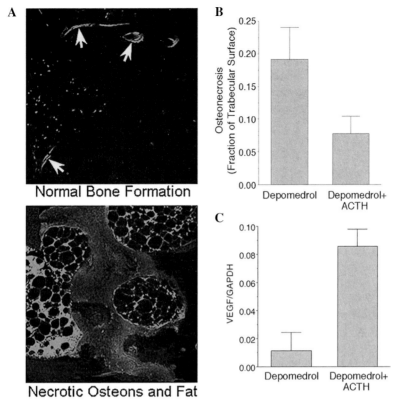

Figure 2. Administration of depot prednisolone to rabbits causes osteonecrosis of trabecular bone of the femoral head, as shown by tetracycline labeling, but ACTH coadministration rescues the bone. (A) In necrotic bone (lower frame) the entire thickness of affected bone is labeled passively.[17] Active tetracycline labeling during bone formation (upper frame) is much more intense and occurs only in living bone (arrows). In necrosis, the marrow accumulates calcium due to necrotic fat and hence also binds tetracycline (bottom). Fields are 220 μm^2. (B) Quantitative analysis showed that animals receiving both depot prednisolone and ACTH had dramatically reduced necrotic trabecular bone surface. The study used year-old rabbits treated 25 days with high-dose depot methylprednisolone (Depomedrol) or the same with low-dose ACTH rescue daily.[22] $n = 4, P < 0.05$. (C) The improved bone survival when ACTH was co-administered with prednisolone correlated with dramatically increased VEGF mRNA production in rabbit femoral head bone cells. Marrow from the animals in (B) was analyzed using quantitative PCR of total bone-derived mRNA. $n = 4, P < 0.01$.

coordinate functions of multiple GPCRs in bone. The FSH is usually balanced by partial, but significant, estrogen effects that reduce osteoclast differentiation, largely by nongenomic mechanisms.[28,29]

Although the physiological importance of low-level expression can always be questioned, we very consistently find FSH-R in human CD14 cells and osteoclasts.[25] We use nested primers and sequencing to verify the specificity of the reactions, and amplify regions that contain an intron to avoid the pitfall of genomic DNA contamination. In keeping with the observed FSH activation of bone loss in mouse and human cells,[26] there are many well-characterized examples where very few GPCRs can activate cellular processes under specific circumstances, including the function of GPCRs in sensory organs. The

bigger issue seems to be why such an effect, which seems to be a paradoxical survival disadvantage, in menopause should occur, that is, low estrogen and high FSH, both favoring bone resorption.[26–29] To some extent it might be expected that paradoxical responses do occur in circumstances, such as survival at an older age, which were rarely achieved even a century ago. But this particular response to ovarian failure is probably protected evolutionarily because it is physiologically important and logical in another circumstance. Specifically, FSH rises to a fairly high plateau, typically over 10 mIU/mL and sometimes exceeding 20 mIU/mL, for prolonged periods during lactation,[30] when estrogen is also low. Often below normal levels for cycling women, but usually not as low as in the menopause. In lactation, thus, FSH would promote availability of calcium, and in this physiologically important circumstance, the elevated FSH with low estrogen causes, to a lesser extent than in menopause but in a similar manner, calcium mobilization in an unbalanced fashion clearly in a survival-related function. Greatly reduced infant mortality and other factors have reduced the awareness of physiological factors relating to lactation, but they are nonetheless vital in the larger context of endocrine metabolic regulation. It should be mentioned in this context that oxytocin is an anabolic bone hormone,[4] in the context of lactation in part opposing the severe skeletal damage that occurs in the menopause.

There were, mostly in the 1970s and 1980s, several reports of FSH-like activity during pregnancy, sometimes attributed to cross-activation of FSH receptors by high levels of human hCG. This has not been demonstrated to occur in humans to the satisfaction of most endocrinologists,[31] although point mutations that produce promiscuous activation are well known, and it is possible that, in the absence of mutations, heterodimers of LH and FSH receptors might produce real, although low-level, cross-activation.[32] Such effects are likely to have no discernible effect except at very high-ligand concentrations, and, then, might be laboratory curiosities. On the other hand, it is not known if FSH receptor variants, such as those in human monocytes,[25] might have altered ligand specificity.

Pituitary glycoproteins are produced outside of the pituitary

Expression of many ligands for pituitary GPCRs have been documented to occur outside the pituitary, typically in specialized cells and at expression levels unlikely to affect circulating hormone levels. This specialized expression would thus be of no importance to regulation of the endocrine organs directly by the pituitary, except in circumstances such as malignancy where large amounts of ectopic hormone production may interfere with the normal pituitary–endocrine axis. But specialized ligand production might affect receptors that occur in peripheral locations, when the producing cells are in proximity to the receptors.

Because POMC is processed to produce several products other than ACTH, the potential for ACTH to be produced widely is apparent. Normally, however, cells that produce other melanocortins do

so with high specificity due to a number of factors including modification of the processing enzymes that channel the products to those specific to individual sites.[33] On the other hand, there are reports of ACTH production by human macrophage/ monocyte cells,[34] so it is possible that some stimulation of ACTH receptors may occur in bone independently of pituitary ACTH. We have been unable to detect significant POMC mRNA in human macrophages (unpublished data), but such production might be context dependent, so it would be best not to regard the subject as closed.

Low-level TSHβ-chain production by bone marrow cells is reported,[35] as a splice variant that is reported to activate the TSH-receptor. This is consistent with other reports of TSH production by lymphocytes.[36,37] This production would be unlikely to be sufficient to affect circulating TSH measurably, but, as with the possibility of local activation of skeletal receptors by ACTH, it might be significant as a limited local signal. This would be similar to the GPCR releasing factors produced in the hypothalamus, at levels generally too low to accumulate in general circulation, but controlling the pituitary due to a special portal vascular system.

In the case of the FSHβ chain, there is no strong evidence of bone or marrow cell production, although coproduction of TSHβ and FSHβ has been reported in mouse CD11β cells from the thyroid.[38] Thus, it is possible that FSH also might be produced in limited quantities near peripheral cells bearing the receptor. The evidence in this case cannot be seen as compelling.

Conflicts of interest

The authors declare no conflict of interest.

References

1. Blair, H.C., A. Wells & C.M. Isales. 2007. Pituitary glycoprotein hormone receptors in non-endocrine organs. *Trends Endocrinol. Metab.* **18:** 227–233.
2. Spudich, J.L., C.S. Yang, K.H. Jung & E.N. Spudich. 2000. Retinylidene proteins: structures and functions from archaea to humans. *Annu. Rev. Cell Dev. Biol.* **16:** 365–392.
3. Ono, N., K. Nakashima, E. Schipani, *et al.* 2012. Constitutively active PTH/PTHRP receptor specifically expressed in osteoblasts enhances bone formation induced by bone marrow ablation. *J. Cell. Physiol.* **227.** [Epub ahead of print]. doi: 10.1002/jcp.22986.
4. Tamma, R., G. Colaianni, L.L. Zhu, *et al.* 2009. Oxytocin is an anabolic bone hormone. *Proc. Natl. Acad. Sci. U.S.A.* **106:** 7149–7154.

5. Albers, J., J. Schulze, F.T. Beil, *et al.* 2011. Control of bone formation by the serpentine receptor Frizzled-9. *J. Cell Biol.* **192:** 1057–1072.
6. Cooray, S.N. & A.J. Clark. 2011. Melanocortin receptors and their accessory proteins. *Mol. Cell. Endocrinol.* **331:** 215–221.
7. Bicknell, A.B. 2008. The tissue-specific processing of pro-opiomelanocortin. *J. Neuroendocrinol.* **20:** 692–699.
8. Tanaka, S. 2003. Comparative aspects of intracellular proteolytic processing of peptide hormone precursors: studies of proopiomelanocortin processing. *Zoolog. Sci.* **20:** 1183–1198.
9. Sirianni, R., K.S. Rehman, B.R. Carr, *et al.* 2005. Corticotropin-releasing hormone directly stimulates cortisol and the cortisol biosynthetic pathway in human fetal adrenal cells. *J. Clin. Endocrinol. Metab.* **90:** 279–285.
10. Vibede, N., F. Hauser, M. Williamson & C.J. Grimmelikhuijzen. 1998. Genomic organization of a receptor from sea anemones, structurally and evolutionarily related to glycoprotein hormone receptors from mammals. *Biochem. Biophys. Res. Commun.* **252:** 497–501.
11. Kumar, R.S., S. Ijiri, K. Kight, *et al.* 2000. Cloning and functional expression of a thyrotropin receptor from the gonads of a vertebrate bony fish: potential thyroid-independent role for thyrotropin in reproduction. *Mol. Cell. Endocrinol.* **167:** 1–9.
12. Kobayashi, T. & O. Andersen. 2008. The gonadotropin receptors FSH-R and LH-R of Atlantic halibut (*Hippoglossus hippoglossus*): isolation of multiple transcripts encoding full-length and truncated variants of FSH-R. *Gen. Comp. Endocrinol.* **156:** 584–594.
13. Bogerd, J., J.C. Granneman, R.W. Schulz & H.F. Vischer. 2005. Fish FSH receptors bind LH: how to make the human FSH receptor to be more fishy? *Gen. Comp. Endocrinol.* **142:** 34–43.
14. Kumar, R.S., S. Ijiri & J.M. Trant. 2001. Molecular biology of the channel catfish gonadotropin receptors: 2. Complementary DNA cloning, functional expression, and seasonal gene expression of the follicle-stimulating hormone receptor. *Biol. Reprod.* **65:** 710–717.
15. Weinstein, R.S. 2010. Glucocorticoids, osteocytes, and skeletal fragility: the role of bone vascularity. *Bone* **46:** 564–570.
16. Mankin, H.J. 1992. Nontraumatic necrosis of bone (osteonecrosis). *N. Engl. J. Med.* **326:** 1473–1479.
17. Eberhardt, A.W., A. Yeager-Jones & H.C. Blair. 2001. Regional trabecular bone matrix degeneration and osteocyte death in femora of glucocorticoid-treated rabbits. *Endocrinology* **142:** 1333–1340.
18. Zhong, Q., S. Sridhar, L. Ruan, *et al.* 2005. Multiple melanocortin receptors are expressed in bone cells. *Bone* **36:** 820–831.
19. Zaidi, M., L. Sun, L.J. Robinson, *et al.* 2010. ACTH protects against glucocorticoid-induced osteonecrosis of bone. *Proc. Natl. Acad. Sci. U.S.A.* **107:** 8782–8787.
20. Wang, G., C.Q. Zhang, Y. Sun, *et al.* 2010. Changes in femoral head blood supply and vascular endothelial growth factor in rabbits with steroid-induced osteonecrosis. *J. Int. Med. Res.* **38:** 1060–1069.
21. Marians, R.C., L. Ng, H.C. Blair, *et al.* 2001. Defining thyrotropin-dependent and -independent steps of thyroid hormone synthesis by using thyrotropin receptor-null mice. *Proc. Natl. Acad. Sci. U.S.A.* **99:** 15776–15781.
22. Abe, E., R.C. Marians, W. Yu, *et al.* 2003. TSH is a negative regulator of skeletal remodeling. *Cell* **115:** 151–162.
23. Hase, H., T. Ando, L. Eldeiry, *et al.* 2006. TNFα mediates the skeletal effects of thyroid-stimulating hormone. *Proc. Natl. Acad. Sci. U.S.A.* **103:** 12849–12854.
24. Wang, H.C., J. Dragoo, Q. Zhou & J.R. Klein. 2003. An intrinsic thyrotropin-mediated pathway of TNF-α production by bone marrow cells. *Blood* **101:** 119–123.
25. Robinson, L.J., I. Tourkova, Y. Wang, *et al.* 2010. FSH-receptor isoforms and FSH-dependent gene transcription in human monocytes and osteoclasts. *Biochem. Biophys. Res. Commun.* **394:** 12–17.
26. Sun, L., Y. Peng, A.C. Sharrow, *et al.* 2006. FSH directly regulates bone mass. *Cell* **125:** 247–260.
27. Iqbal, J., L. Sun, T.R. Kumar, *et al.* 2006. Follicle-stimulating hormone stimulates TNF production from immune cells to enhance osteoblast and osteoclast formation. *Proc. Natl. Acad. Sci. U.S.A.* **103:** 14925–14930.
28. García Palacios, V., L.J. Robinson, C.W. Borysenko, *et al.* 2005. Negative regulation of RANKL-induced osteoclastic differentiation in RAW264.7 Cells by estrogen and phytoestrogens. *J. Biol. Chem.* **280:** 13720–13727.
29. Robinson, L.J., B.B. Yaroslavskiy, R.D. Griswold, *et al.* 2009. Estrogen inhibits RANKL-stimulated osteoclastic differentiation of human monocytes through estrogen and RANKL-regulated interaction of estrogen receptor-α with BCAR1 and Traf6. *Exp. Cell Res.* **315:** 1287–1301.
30. Burger, H.G., J.P. Hee, P. Mamers, *et al.* 1994. Serum inhibin during lactation: relation to the gonadotrophins and gonadal steroids. *Clin. Endocrinol. (Oxford).* **41:** 771–777.
31. Simoni, M., S.A. Khan & E. Nieschlag. 1991. Serum bioactive follicle-stimulating hormone-like activity in human pregnancy is a methodological artifact. *J. Clin. Endocrinol. Metab.* **73:** 1118–1122.
32. Costagliola, S., E. Urizar, F. Mendive & G. Vassart. 2005. Specificity and promiscuity of gonadotropin receptors. *Reproduction* **130:** 275–281.
33. Bruzzaniti, A., R. Marx & R.E. Mains. 1999. Activation and routing of membrane-tethered prohormone convertases 1 and 2. *J. Biol. Chem.* **274:** 24703–24713.
34. Pállinger, E. & G. Csaba. 2008. A hormone map of human immune cells showing the presence of adrenocorticotropic hormone, triiodothyronine and endorphin in immunophenotyped white blood cells. *Immunology* **123:** 584–589.
35. Vincent, B.H., D. Montufar-Solis, B.B. Teng, *et al.* 2009. Bone marrow cells produce a novel TSHβ splice variant that is upregulated in the thyroid following systemic virus infection. *Genes Immun.* **10:** 18–26.
36. Smith, E.M., M. Phan, T.E. Kruger, *et al.* 1983. Human lymphocyte production of immunoreactive thyrotropin. *Proc. Natl. Acad. Sci. U.S.A.* **80:** 6010–6013.
37. Harbour, D.V., T.E. Kruger, D. Coppenhaver, *et al.* 1989. Differential expression and regulation of thyrotropin (TSH) in T cell lines. *Mol. Cell. Endocrinol.* **64:** 229–241.
38. Klein, J.R. & H.C. Wang. 2004. Characterization of a novel set of resident intrathyroidal bone marrow-derived hematopoietic cells: potential for immune-endocrine interactions in thyroid homeostasis. *J. Exp. Biol.* **207**(Pt. 1): 55–65.

Ann. N.Y. Acad. Sci. ISSN 0077-8923

ANNALS OF THE NEW YORK ACADEMY OF SCIENCES
Issue: *Skeletal Biology and Medicine II*

Mechanotransduction and cartilage integrity

Daniel J. Leong,[1,2,3] John A. Hardin,[1] Neil J. Cobelli,[1] and Hui B. Sun[1,2,3]

[1]Department of Orthopaedic Surgery, [2]Department of Radation Oncology, [3]Oncophysics Research Institute, Albert Einstein College of Medicine/Montefiore Medical Center, Bronx, New York

Address for correspondence: Hui B. Sun, Ph.D., Department of Orthopaedic Surgery and Radiation Oncology, Albert Einstein College of Medicine and Montefiore Medical Center, 1300 Morris Park Avenue, Golding Building Room 103, Bronx, NY 10461. Herb.sun@einstein.yu.edu

Osteoarthritis (OA) is characterized by the breakdown of articular cartilage that is mediated in part by increased production of matrix metalloproteinases (MMPs) and aggrecanases (ADAMTS), enzymes that degrade components of the cartilage extracellular matrix. Efforts to design synthetic inhibitors of MMPs/ADAMTS have only led to limited clinical success. In addition to pharmacologic therapies, physiologic joint loading is widely recommended as a nonpharmacologic approach to improve joint function in osteoarthritis. Clinical trials report that moderate levels of exercise exert beneficial effects, such as improvements in pain and physical function. Experimental studies demonstrate that mechanical loading mitigates joint destruction through the downregulation of MMPs/ADAMTS. However, the molecular mechanisms underlying these effects of physiologic loading on arthritic joints are not well understood. We review here the recent progress on mechanotransduction in articular joints, highlighting the mediators and pathways in the maintenance of cartilage integrity, especially in the prevention of cartilage degradation in OA.

Keywords: mechanical loading; osteoarthritis; exercise; cartilage degradation

Introduction

Osteoarthritis (OA) is a progressive degenerate joint disease that affects the structural and functional integrity of joint tissues such as bone, tendons, and ligaments, which ultimately results in the destruction of articular cartilage. It is currently the leading cause of disability and pain in the United States,[1] and there are currently no cures for OA, and no effective pharmacological treatments exist that slow or halt its progression.[2] Physical activity is one of the most widely prescribed nonpharmacological therapies for OA management,[3] based on its ability to limit pain and improve physical function.[4,5] However, the mechanisms underlying these beneficial effects of exercise and physical therapy (referred in this paper as "mechanical treatment") are largely unknown. In this review, we will discuss the recent progress regarding the effects of mechanical treatment on OA, and highlight a novel mechanotransduction pathway that mediates the anti-inflammatory and chondroprotective effects of physiologic joint loading.

Cartilage destruction in osteoarthritis

Osteoarthritis is characterized by cartilage degradation, synovial inflammation, and alterations within the subchondral bone, including bone remodeling, subchondral sclerosis, and osteophyte formation.[2,6] Clinical features of osteoarthritis include joint pain, stiffness, and swelling, which together contribute to patient disability.[7] The pathogenesis of OA is unclear, but risk factors for developing OA include aging, joint trauma, obesity, and heritable genetic factors.[2] OA is the most common joint disease, affecting an estimated 15% of the U.S. population.[8]

At the molecular level, one of the most prominent features of OA is the imbalance between the anabolic and catabolic activities within chondrocytes, the sole cell population within cartilage. Breakdown of the cartilage extracellular matrix is mediated in part

doi: 10.1111/j.1749-6632.2011.06301.x

by upregulated expression of proteolytic enzymes, including matrix metalloproteinases (MMPs) or a disintegrin and metalloproteinase with thrombospondin motifs (ADAMTS).[9] In cases of normal tissue turnover, levels of active MMPs and ADAMTS are suppressed, in part by tissue inhibitors of metalloproteinases (TIMPs).[10] However, in osteoarthritis, the activities of proteolytic enzymes overwhelm those of TIMPs, resulting in cartilage breakdown.[2]

Due to the role upregulated MMPs/ADAMTS play in arthritis, inhibitors for these proteolytic enzymes have been explored as therapeutic strategies to treat OA. However, clinical trials so far have been met with limited success and resulted in side effects including musculoskeletal pain and inflammation.[11–13] These adverse effects have been mainly attributed to the lack of selectivity of these inhibitors. Metalloproteinases share structural similarities and are susceptible to regulation by broad-spectrum inhibitors.[14] Poor selectivity is problematic because in addition to matrix remodeling, MMPs/ADAMTS play important roles in wound healing, angiogenesis, development, morphogenesis, and bone remodeling.[15,16] Therefore, it appears successful therapeutic strategies will require the specific inhibition and appropriate modulation of MMPs/ADAMTS involved in OA.

Physiologic joint loading and osteoarthritis

Nonpharmacologic therapies for OA, such as aerobic exercise, strength training, and passive motion therapy, have been reported to exert protective effects on the joint. At least 20 min of weekly rigorous physical exercise, defined as activities leading to shortness of breath or sweating, is protective against the development of cartilage defects in healthy adults.[17,18] Less vigorous physical activities such as walking are also beneficial for joint health. Subjects who walk regularly (more than three times a week for at least 20 min each time) have a reduced risk of developing bone marrow lesions.[17] Bone marrow lesions are associated with the development of chondral defects and may serve as a predictive biomarker of OA development.[19]

For people with OA, regular exercise also has been demonstrated to be of benefit. A Cochrane Review of 32 clinical trials comparing land-based therapeutic exercise (i.e., muscle strengthening, aerobics, manual therapy) to a nonexercise group found that exercise treatment resulted in moderate improvements in pain and physical function.[20] Physical interventions that are less studied, including hydrotherapy and Tai Chi, have reported significant improvements in pain and physical function for at least 24 weeks after the start of these exercise programs.[21] Although clinical trials evaluating the effect of exercise on joint structure in OA patients are limited, preliminary results from these studies are promising. A four-month exercise program consisting of aerobic and weight-bearing exercises increased proteoglycan content in the articular cartilage of OA subjects.[22] Strength training, when compared to range of motion exercises for 30 months, decreased the mean rate of joint space narrowing, but the difference was not statistically significant.[23] Together, the evidence support that moderate levels of exercise improve symptoms of OA, but whether exercise also has a disease modifying effect is still unclear.

Molecular effects of exercise in the synovial joint

Although the beneficial effects of exercise for osteoarthritis patients are well documented, the mechanisms are still largely unknown. *In vitro* and *in vivo* experiments have begun to identify the molecular effects of physiologic loading in chondrocytes, and determine factors mediating the beneficial actions of loading. *In vitro*, proinflammatory cytokine interleukin (IL)-1β stimulates the release of proinflammatory mediators such as nitric oxide (NO), prostaglandin (PG)E_2, and cyclo-oxygenase (COX)-2.[24] Dynamic compression of chondrocytes at 0–15% strain and 1 hertz (Hz) counteracts the production of these IL-1β–induced mediators, possibly through a p38 mitogen-activated protein kinase (MAPK)-dependent pathway.[24] Mechanical stimulation of chondrocytes also antagonizes IL-1β– and TNF-α–induced inflammatory and catabolic responses, such as upregulated COX-2, inducible nitric oxide synthase (iNOS), and genes involved in cartilage catabolism, such as MMPs 9 and 13.[25–27] This beneficial effect of mechanical loading was attributed to inhibiting transcription factor nuclear factor-kappa B (NF-κB) from translocating into the nucleus to activate target genes and also to interference with multiple signaling events upstream of NF-κB.[28] Another mechanism through which mechanical loading acts is by preventing IL-1α–induced cartilage degradation.[29] Bovine cartilage explants incubated with IL-1α led to the degradation of collagen and proteoglycans and resulted in aggrecan cleavage by MMPs and ADAMTS. Dynamic loading at 0.5 megapascals (MPa) and 0.5Hz of these explants inhibited the catabolic actions of IL-1α and prevented cartilage degradation.[29]

In vivo experiments have clearly demonstrated the anti-catabolic and anti-inflammatory effects of physiologic joint loading. Hindlimb immobilization of rodents resulted in catabolic changes, including reduced Safranin O staining, indicative of proteoglycan loss, and increases in MMP-3 and ADAMTS-5.[30] However, 1 h of daily passive joint motion inhibited the increases in MMP-3 and ADAMTS-5 and prevented changes in proteoglycan loss.[30] In animal models of antigen-induced arthritis, daily bouts of passive motion therapy decreased joint inflammation and maintained the structural integrity of the articular cartilage when compared to immobilized controls, demonstrating its potential for therapeutic use. Mechanistically, passive motion therapy exerted potent anti-inflammatory effects. Passive motion significantly decreased the levels of proinflammatory genes and mediators of matrix breakdown (IL-1β, COX-2, MMP-1) and induced anti-inflammatory cytokine IL-10.[26,31] IL-10 has protective effects in cartilage,[32] and its induction may be one mechanism by which mechanical signals render anti-inflammatory effects. Together, the *in vitro* and *in vivo* data suggest that a variety of loading conditions are sufficient to preserve cartilage integrity by counteracting cytokine-induced proinflammatory and catabolic effects.

In addition to the direct effects mechanical loading exerts on chondrocytes, exercise can affect the synovial cavity. One bout of exercise in female patients with OA increased the concentration of IL-10 in the synovial fluid and in the peri-synovial compartment when compared to a non-exercise group.[33] Passive mobilization of knee joints in anesthetized rabbits increased hyaluronan (HA) secretion when compared to static controls.[34] Hyaluronan is synthesized by synoviocytes and contributes to the lubricating capacity of synovial fluid.[35] In patients with OA, the concentration of HA is reduced,[36] and intra-articular injections of HA are widely used for the relief of knee pain associated with OA.[37]

Chondrocyte mechanotransduction

Mechanotransduction is the process by which biomechanical signals regulate cell activity and behavior. Chondrocytes are able to sense and react to mechanically induced changes within the cartilage matrix.[38] Chondrocyte mechanotransduction is initiated at the interface between the cell membrane and extracellular matrix,[39] and the processing of these mechanical signals involves mechanoreceptors such as ion channels and integrins. For example, membrane stretch, a condition that chondrocytes experience during compression or during hypo-osmotic conditions that cause swelling,[40] activates potassium channels.[41] The function of ion channels in chondrocyte membranes is not clear, but they may be involved in chondrocyte functions such as cell proliferation and matrix secretion.[42,43] Integrins are heterodimeric transmembrane receptors consisting of α and β subunits[44] and interact with cytoskeletal proteins such as fibronectin, vitronectin, and osteopontin.[45–47] Mechanical stimulation of human chondrocytes increases expression of aggrecan and decreases MMP-3 gene expression in a pathway involving the $α_5β_1$ integrin and IL-4 release.[48] However, this response to mechanical stimulation is absent in chondrocytes derived from OA cartilage, suggesting abnormal chondrocyte signaling may be involved in OA disease progression.[49]

Little is known of the joint loading-activated signaling pathways that help maintain cartilage integrity in OA. To identify the molecular basis of exercise in osteoarthritis, transcriptome-wide microarray analysis was performed in rodents experimentally induced with arthritis and either run on a treadmill daily for 21 days or subject to cage activity. Treadmill exercise initiated one day after arthritis induction significantly slowed progression of arthritis, while upregulating gene networks associated with matrix synthesis and suppressing proinflammatory gene networks.[50] Of interest, treadmill exercise initiated five or nine days after arthritis induction, when cartilage destruction was more severe, was less effective in protecting articular cartilage from destruction.[50]

The NF-κB network was one of the gene clusters suppressed by treadmill exercise.[50] NF-κB transcription factors are involved in immune and inflammatory responses and regulate expression of genes responsible for inflammation, apoptosis, cell cycle, and matrix breakdown.[51,52] In response to various stimuli such as TNF-α, IL-1β, and lipopolysaccharides (LPS), NF-κB is activated and translocates to the nucleus to regulate transcription of its target genes.[53] With treadmill exercise, expression of many genes required for NF-κB activity was suppressed, suggesting the suppression of NF-κB activation mediates the anti-inflammatory effects of exercise.[50]

CITED2-mediated mechanotransduction

One transcriptional regulator that appears to play a crucial role in cartilage homeostasis is CITED2 (CBP/p300-interacting transactivator with ED-rich tail 2). CITED2 is a transcriptional coregulator that does not bind DNA directly. It positively regulates transcription by recruiting CBP (cAMP-responsive element-binding protein) and p300 to interact with other DNA-binding transcription factors such as Lhx2, PPARα, PPARγ, Smad 2, and TFAP2.[54] CITED2 also negatively regulates target genes by competing for CBP/p300 binding with transcription factors including Ets-1, NF-κB, HIF-1α, STAT2, and p53.[55,56] Through these mechanisms, CITED2 is able to regulate many cellular processes such as embryonic development, cell proliferation, inflammation, and matrix turnover.[54]

With regard to cartilage integrity, CITED2 expression in chondrocytes *in vitro* is increased by moderate intensities of flow shear and intermittent hydrostatic pressure (IHP) and in chondrocytes *in vivo* by joint motion.[57,58] Increased CITED2 expression *in vivo* correlated with the maintenance of cartilage integrity and the suppression of collagenase MMP-1, suggesting the anticatabolic effects of physiologic joint loading were mediated by CITED2.[58] As demonstrated by competitive binding and transcriptional activity assays, CITED2 suppresses MMP-1 transcription by competing with MMP transactivator Ets-1 for binding to its coactivator p300. In addition to MMP-1, Ets-1 binds to the promoter regions of other MMPs including MMP-2, -3, -8, -9, and -13.[59] Therefore, it is likely CITED2 may regulate additional MMPs through a similar manner.

Upstream of CITED2, moderate IHP loading phosphorylated p38δ, which was required for the transactivation of CITED2.[58] p38 belongs to the MAP kinase family, which is activated in response to mechanical stresses.[60] While moderate loading ac-

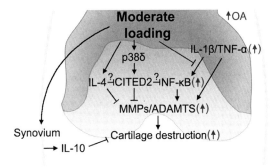

Figure 1. CITED2 as a central mediator of the hypothesized mechanotransduction pathways in the maintenance of cartilage integrity in healthy joints by balancing catabolic and anticatabolic events, and in the reduction or prevention of cartilage degradation in diseased (OA) joints by suppressing upregulated proinflammatory networks in OA.

tivated p38δ and CITED2, high levels of IHP phorphorylated p38α and MMP-1, but not CITED2. This may explain why CITED2 is specifically activated by moderate loading and also suggests that p38α is involved in the upregulation of MMP-1. The data indicate different members of the p38 family may act as a "mechanosensitive switch" in chondrocytes, which act to upregulate or downregulate MMP expression based on the mechanical loading regimes. The evidence that CITED2 is inducible by IL-4[61] and may interact with components of NF-κB,[56] suggests a potential role of CITED2 as a central mediator in these mechanotransduction pathways involved in maintaining cartilage integrity (Fig. 1).

Conclusions

Physical activity is widely prescribed as a non-pharmacologic therapy for patients with OA. While moderate levels of physical activity are reported to exert beneficial clinical effects in patients with OA, the parameters for each type of exercise (i.e., intensity, duration, frequency) are not well characterized within the OA population. Furthermore, the mechanisms of how mechanical signals mediate these effects are not well understood. The recent identification of mechanotransduction pathways in response to physiological joint loading, such as the CITED2-mediated pathway, and its potential cross-talk to pathways mediated by NF-κB, contribute to our understanding of mechanisms underlying mechanical treatment, and may lead to novel therapeutic targets and strategies to treat or prevent cartilage destruction in arthritis.

Acknowledgements

The work has been supported by NIH Grants AR47628 and AR52743.

Conflicts of interest

The authors declare no conflicts of interest.

References

1. Lawrence, R.C. *et al.* 2008. Estimates of the prevalence of arthritis and other rheumatic conditions in the United States: part II. *Arthritis Rheum.* **58:** 26–35.
2. Sun, H.B. 2010. Mechanical loading, cartilage degradation, and arthritis. *Ann. N. Y. Acad. Sci.* **1211:** 37–50.
3. Ng, N.T., K.C. Heesch & W.J. Brown. 2011. Strategies for managing osteoarthritis. *Int. J. Behav. Med.* May 26. [Epub ahead of print].
4. Fransen, M., S. McConnell & M. Bell. 2002. Therapeutic exercise for people with osteoarthritis of the hip or knee. A systematic review. *J. Rheumatol.* **29:** 1737–1745.
5. Roddy, E., W. Zhang & M. Doherty. 2005. Aerobic walking or strengthening exercise for osteoarthritis of the knee? A systematic review. *Ann. Rheum. Dis.* **64:** 544–548.
6. Yokota, H., D.J. Leong & H.B. Sun. 2011. Mechanical loading: bone remodeling and cartilage maintenance. *Curr. Osteoporos. Rep.* **9:** 237–242.
7. Bijlsma, J.W., F. Berenbaum & F.P. Lafeber. 2011. Osteoarthritis: an update with relevance for clinical practice. *Lancet* **377:** 2115–2126.
8. Felson, D.T. *et al.* 2000. Osteoarthritis: new insights. Part 1: the disease and its risk factors. *Ann. Intern. Med.* **133:** 635–646.
9. Goldring, M.B. & K.B. Marcu. 2009. Cartilage homeostasis in health and rheumatic diseases. *Arthritis Res. Ther.* **11:** 224–240.
10. Brew, K. & H. Nagase. 2010. The tissue inhibitors of metalloproteinases (TIMPs): an ancient family with structural and functional diversity. *Biochim. Biophys. Acta* **1803:** 55–71.
11. Coussens, L.M., B. Fingleton L.M. Matrisian. 2002. Matrix metalloproteinase inhibitors and cancer: trials and tribulations. *Science* **295:** 2387–2392.
12. Devy, L. & D.T. Dransfield. 2011. New strategies for the next generation of matrix-metalloproteinase inhibitors: selectively targeting membrane-anchored MMPs with therapeutic antibodies. *Biochem. Res. Int.* **2011:** 191670–191681.
13. Skiles, J.W., N.C. Gonnella & A.Y. Jeng. 2004. The design, structure, and clinical update of small molecular weight matrix metalloproteinase inhibitors. *Curr. Med. Chem.* **11:** 2911–2977.
14. Fanjul-Fernandez, M., A.R. Folgueras, S. Cabrera & C. Lopez-Otin. 2010. Matrix metalloproteinases: evolution, gene regulation and functional analysis in mouse models. *Biochim. Biophys. Acta* **1803:** 3–19.
15. McQuibban, G.A. *et al.* 2002. Matrix metalloproteinase processing of monocyte chemoattractant proteins generates CC chemokine receptor antagonists with anti-inflammatory properties in vivo. *Blood* **100:** 1160–1167.
16. Nagase, H., R. Visse & G. Murphy. 2006. Structure and function of matrix metalloproteinases and TIMPs. *Cardiovasc. Res.* **69:** 562–573.
17. Racunica, T.L. *et al.* 2007. Effect of physical activity on articular knee joint structures in community-based adults. *Arthritis Rheum.* **57:** 1261–1268.
18. Foley, S., C. Ding, F. Cicuttini & G. Jones. 2007. Physical activity and knee structural change: a longitudinal study using MRI. *Med. Sci. Sports Exerc.* **39:** 426–434.
19. Wluka, A.E. *et al.* 2009. Bone marrow lesions predict increase in knee cartilage defects and loss of cartilage volume in middle-aged women without knee pain over 2 years. *Ann. Rheum. Dis.* **68:** 850–855.
20. Fransen, M. & S. McConnell. 2008. Exercise for osteoarthritis of the knee. *Cochrane Database Syst. Rev.* **8:** CD004376-CD004468.
21. Fransen, M., L. Nairn, J. Winstanley, *et al.* 2007. Physical activity for osteoarthritis management: a randomized controlled clinical trial evaluating hydrotherapy or Tai Chi classes. *Arthritis Rheum.* **57:** 407–414.
22. Roos, E.M. & L. Dahlberg. 2005. Positive effects of moderate exercise on glycosaminoglycan content in knee cartilage: a four-month, randomized, controlled trial in patients at risk of osteoarthritis. *Arthritis Rheum.* **52:** 3507–3514.
23. Mikesky, A.E. *et al.* 2006. Effects of strength training on the incidence and progression of knee osteoarthritis. *Arthritis Rheum.* **55:** 690–699.
24. Chowdhury, T.T. *et al.* 2008. Dynamic compression counteracts IL-1beta induced inducible nitric oxide synthase and cyclo-oxygenase-2 expression in chondrocyte/agarose constructs. *Arthritis Res. Ther.* **10:** R35–R48.
25. Gassner, R. *et al.* 1999. Cyclic tensile stress exerts antiinflammatory actions on chondrocytes by inhibiting inducible nitric oxide synthase. *J. Immunol* **163:** 2187–2192.
26. Ferretti, M. *et al.* 2006. Biomechanical signals suppress proinflammatory responses in cartilage: early events in experimental antigen-induced arthritis. *J. Immunol.* **177:** 8757–8766.
27. Madhavan, S. *et al.* 2006. Biomechanical signals exert sustained attenuation of proinflammatory gene induction in articular chondrocytes. *Osteoarthritis Cartilage* **14:** 1023–1032.
28. Dossumbekova, A. *et al.* 2007. Biomechanical signals inhibit IKK activity to attenuate NF-kappaB transcription activity in inflamed chondrocytes. *Arthritis Rheum.* **56:** 3284–3296.
29. Torzilli, P.A., M. Bhargava, S. Park & C.T. Chen. 2010. Mechanical load inhibits IL-1 induced matrix degradation in articular cartilage. *Osteoarthritis Cartilage* **18:** 97–105.
30. Leong, D.J. *et al.* 2010. Matrix metalloproteinase-3 in articular cartilage is upregulated by joint immobilization and suppressed by passive joint motion. *Matrix Biol.* **29:** 420–426.
31. Ferretti, M. *et al.* 2005. Anti-inflammatory effects of continuous passive motion on meniscal fibrocartilage. *J. Orthop. Res.* **23:** 1165–1171.
32. Lubberts, E. *et al.* 2000. Intra-articular IL-10 gene transfer regulates the expression of collagen-induced arthritis (CIA) in the knee and ipsilateral paw. *Clin. Exp. Immunol.* **120:** 375–383.

33. Helmark, I.C. *et al.* 2010. Exercise increases interleukin-10 levels both intraarticularly and peri-synovially in patients with knee osteoarthritis: a randomized controlled trial. *Arthritis Res. Ther.* **12:** R126–R137.

34. Ingram, K.R., A.K. Wann, C.K. Angel, *et al.* 2008. Cyclic movement stimulates hyaluronan secretion into the synovial cavity of rabbit joints. *J. Physiol.* **586:** 1715–1729.

35. Campbell, J., N. Bellamy & T. Gee. 2007. Differences between systematic reviews/meta-analyses of hyaluronic acid/hyaluronan/hylan in osteoarthritis of the knee. *Osteoarthritis Cartilage* **15:** 1424–1436.

36. Altman, R.D. 2003. Status of hyaluronan supplementation therapy in osteoarthritis. *Curr. Rheumatol. Rep.* **5:** 7–14.

37. Barron, M.C. & B.R. Rubin. 2007. Managing osteoarthritic knee pain. *J. Am. Osteopath. Assoc.* **107:** ES21–ES27.

38. Ramage, L., G. Nuki & D.M. Salter. 2009. Signalling cascades in mechanotransduction: cell-matrix interactions and mechanical loading. *Scand. J. Med. Sci. Sports* **19:** 457–469.

39. Wang, N., J.D. Tytell & D.E. Ingber. 2009. Mechanotransduction at a distance: mechanically coupling the extracellular matrix with the nucleus. *Nat. Rev. Mol. Cell Biol.* **10:** 75–82.

40. Urban, J.P., A.C. Hall & K.A. Gehl. 1993. Regulation of matrix synthesis rates by the ionic and osmotic environment of articular chondrocytes. *J. Cell Physiol.* **154:** 262–270.

41. Mobasheri, A. *et al.* 2010. Characterization of a stretch-activated potassium channel in chondrocytes. *J. Cell Physiol.* **223:** 511–518.

42. Mouw, J.K., S.M. Imler & M.E. Levenston. 2007. Ion-channel regulation of chondrocyte matrix synthesis in 3D culture under static and dynamic compression. *Biomech. Model Mechanobiol.* **6:** 33–41.

43. Wu, Q.Q. & Q. Chen. 2000. Mechanoregulation of chondrocyte proliferation, maturation, and hypertrophy: ion-channel dependent transduction of matrix deformation signals. *Exp. Cell Res.* **256:** 383–391.

44. Millward-Sadler, S.J. & D.M. Salter. 2004. Integrin-dependent signal cascades in chondrocyte mechanotransduction. *Ann. Biomed. Eng.* **32:** 435–446.

45. Kock, L.M. *et al.* 2009. RGD-dependent integrins are mechanotransducers in dynamically compressed tissue-engineered cartilage constructs. *J. Biomech.* **42:** 2177–2182.

46. Loeser, R.F. 2002. Integrins and cell signaling in chondrocytes. *Biorheology* **39:** 119–124.

47. Van der Kraan, P.M., P. Buma, T. Van Kuppevelt & W.B. Van den Berg. 2002. Interaction of chondrocytes, extracellular matrix and growth factors: relevance for articular cartilage tissue engineering. *Osteoarthritis Cartilage* **10:** 631–637.

48. Wright, M.O. *et al.* 1997. Hyperpolarisation of cultured human chondrocytes following cyclical pressure-induced strain: evidence of a role for alpha 5 beta 1 integrin as a chondrocyte mechanoreceptor. *J. Orthop. Res.* **15:** 742–747.

49. Millward-Sadler, S.J., M.O. Wright, L.W. Davies, *et al.* 2000. Mechanotransduction via integrins and interleukin-4 results in altered aggrecan and matrix metalloproteinase 3 gene expression in normal, but not osteoarthritic, human articular chondrocytes. *Arthritis Rheum.* **43:** 2091–2099.

50. Nam, J. *et al.* 2011. Transcriptome-wide gene regulation by gentle treadmill walking during the progression of monoiodoacetate-induced arthritis. *Arthritis Rheum.* **63:** 1613–1625.

51. Marcu, K.B., M. Otero, E. Olivotto, *et al.* 2010. NF-kappaB signaling: multiple angles to target OA. *Curr. Drug Targets* **11:** 599–613.

52. Anghelina, M. *et al.* 2008. Regulation of biomechanical signals by NF-kappaB transcription factors in chondrocytes. *Biorheology* **45:** 245–256.

53. Ghosh, S., M.J. May & E.B. Kopp. 1998. NF-kappa B and Rel proteins: evolutionarily conserved mediators of immune responses. *Annu. Rev. Immunol.* **16:** 225–260.

54. Sun, H.B. 2010. CITED2 mechanoregulation of matrix metalloproteinases. *Ann. N. Y. Acad. Sci.* **1192:** 429–436.

55. Dial, R., Z.Y. Sun & S.J. Freedman. 2003. Three conformational states of the p300 CH1 domain define its functional properties. *Biochemistry* **42:** 9937–9945.

56. Lou, X. *et al.* 2011. Negative feedback regulation of NF-kappaB action by CITED2 in the nucleus. *J. Immunol.* **186:** 539–548.

57. Yokota, H., M.B. Goldring & H.B. Sun. 2003. CITED2-mediated regulation of MMP-1 and MMP-13 in human chondrocytes under flow shear. *J. Biol. Chem.* **278:** 47275–47280.

58. Leong, D.J. *et al.* 2011. Physiological loading of joints prevents cartilage degradation through CITED2. *FASEB J.* **25:** 182–191.

59. Dittmer, J. 2003. The biology of the Ets1 proto-oncogene. *Mol. Cancer* **2:** 29–40.

60. Fitzgerald, J.B. *et al.* 2008. Shear- and compression-induced chondrocyte transcription requires MAPK activation in cartilage explants. *J. Biol. Chem.* **283:** 6735–6743.

61. Sun, H.B., Y.X. Zhu, T. Yin, G. Sledge & Y.C. Yang. 1998. MRG1, the product of a melanocyte-specific gene related gene, is a cytokine-inducible transcription factor with transformation activity. *Proc. Natl. Acad. Sci. USA* **95:** 13555–13560.

Ann. N.Y. Acad. Sci. ISSN 0077-8923

ANNALS OF THE NEW YORK ACADEMY OF SCIENCES
Issue: *Skeletal Biology and Medicine II*

Corticosteroid-associated avascular necrosis: dose relationships and early diagnosis

Roy K. Aaron,[1] Anne Voisinet,[1] Jennifer Racine,[1] Yousaf Ali,[2] and Edward R. Feller[3]

[1]Department of Orthopaedics, Warren Alpert Medical School, Brown University, Providence, Rhode Island. [2]Department of Medicine, Mount Sinai School of Medicine, New York, New York. [3]Department of Community Medicine, Warren Alpert Medical School, Brown University, Providence, Rhode Island

Address for correspondence: Roy K. Aaron, M.D. 100 Butler Drive Providence, RI 02906. Roy_Aaron@Brown.edu

Corticosteroids are the most common etiological factor in nontraumatic avascular necrosis (AVN) of bone, accounting for about 10% of arthroplasties performed annually in the United States. Evidence is conflicting on the relative importance of peak dose, daily dose, or cumulative dose, and most likely all three represent "high dose" corticosteroid administration and play a role in AVN. The etiology may be multifactorial with corticosteroids superimposed on genetic or pathological predispositions. Joint preservation depends upon early diagnosis and treatment before fracture of the subchondral trabeculae and joint incongruity. Early intervention depends upon identifying at-risk patients and quantifying their risk by understanding clinical and pathophysiological contributions to that risk. Our data and that of others suggest that a screening MRI of at-risk populations will permit detection of AVN at a prefracture stage when preservation of the joint is possible.

Keywords: Corticosteroids; avascular necrosis; systemic lupus erythematosus; renal transplant

Avascular necrosis (AVN) of bone is a fairly common condition with the incidence varying greatly with the center reporting. AVN accounts for 5–10% of total hip replacements performed in the United States.[1,2] It occurs in various series in 5–25% of patients taking corticosteroids.[1] In Japan, a tenfold increase in the incidence of AVN was noted from 1965 to 1985, and the number of patients diagnosed with AVN doubled between 1984 and 1987.[3] The male-to-female ratio is approximately 4:1, and the mean age of onset is in the fifth decade. AVN has been reported to be bilateral in 50% or more of cases.[4–8] The presentation of symptoms may be asynchronous, but because progression occurs regardless of the temporal appearance of symptoms, a high index of suspicion must be maintained for disease bilaterality. One study reported an incidence of bilaterality, diagnosed by biopsy of 89%.[5] The presence of symptoms in a contralateral hip is not necessary for progression. In a study of asymptomatic contralateral hips in patients with AVN, 100% of the "silent" hips collapsed in a mean of 23 months,

and 83% collapsed in under 36 months. Seventy-nine percent underwent hip replacement in less than three years.[9] AVN is most common in the hip followed by the knee and shoulder. Most attention has been paid to the femoral head because it is the most frequent site of AVN, but concepts of etiology and pathophysiology apply also to other subchondral bone sites.

Although metadiaphyseal infarcts are seen commonly in sickle cell disease and corticosteroid administration, the major significance of AVN is its frequent occurrence in subchondral bone leading to trabecular resorption, insufficiency fractures, joint incongruity, and secondary arthritis requiring joint replacement for palliation. The mean age of patients with AVN is approximately 40 years, an age group in which joint replacement is unlikely to be durable for their life span and has exhibited significant failure rates, often requiring one or more revision arthroplasties.

Joint preservation without arthroplasty depends upon diagnosis and treatment before subchondral

doi: 10.1111/j.1749-6632.2011.06218.x

fracture and collapse of the subchondral bone plate. Regardless of the specific grading system used, subchondral fracture and joint incongruity makes joint preservation unlikely and commits the joint to arthroplasty for pain control, stability, and function. Therefore, the keys to successful joint preservation lie in identifying at-risk patients and quantifying their risk by understanding clinical and pathophysiological contributions to that risk. Appropriate diagnostic techniques may then be applied to at-risk populations for early diagnosis.

Etiologic associations

AVN of the hip occurs frequently after trauma—usually femoral neck fracture or hip dislocation. Several etiologic associations have been proposed for nontraumatic AVN including dysbarism, corticosteroids, alcohol, hemoglobinopathies, Gaucher's disease, and many other less well-documented associations. Methodological problems have made comparison among studies difficult. Impressions of the etiologic relationship between corticosteroids and AVN were initially formed by uncontrolled cross-sectional studies. These studies described the prevalence of corticosteroid use in AVN patients; however, only longitudinal studies of patients taking corticosteroids can determine the prevalence of AVN in corticosteroid-treated patients. Corticosteroid-associated AVN has been observed in various diseases including systemic lupus, rheumatoid arthritis, asthma, inflammatory bowel diseases, and organ transplantation. In addition, studies have variously reported cohort means of the duration of administration, cumulative (total) dose, daily dose, and peak dose. Few studies have estimated the prevalence of AVN in corticosteroid-treated patients with relative risk calculations.

Corticosteroid-associated AVN accounts for approximately 10–30% of nontraumatic AVN, depending upon the center reporting, making it the most common cause of nontraumatic AVN. The incidence of corticosteroid-associated AVN is approximately 10,000–20,000 cases per year in the United States, accounting for approximately 10% of the arthroplasties performed annually.[1]

Although it is not certain that the use of steroids presents equal risks in all clinical conditions, attention has been focused on identifying a threshold dose of steroids in determining risk for AVN. Most studies have suggested an association of AVN in-

cidence with daily or peak dose and have inferred that high doses, even for short duration, present more significant risks than do cumulative dose or duration of therapy. One study of 161 patients with inflammatory bowel disease treated with corticosteroids reported an incidence of AVN of 4.3%. The mean daily prednisone dose was 26 mg/day (range: 12–34 mg/day), the mean duration of treatment was 42 weeks (range: 20–84 weeks), and the mean cumulative prednisone dose was 7,000 mg (range: 1,800–13,500 mg).[10] In a series of 110 patients with steroid-associated AVN with various diseases, the cumulative prednisone dose was 42 g.[11] In heart transplant patients, no association between the development of AVN and cumulative prednisone dose was found, but there was an association between AVN and the peak dose of methylprednisolone.[12] Most patients with AVN in these and other studies in which daily dose was reported received doses of greater than 20 mg/day, and that dose is generally regarded as presenting a threshold risk for AVN.[13–16] A threshold cumulative prednisone dose of 2,000 mg has been reported to present a risk factor for AVN, but there are several reports of patients developing AVN after short-term steroid administration.[17]

The majority of reports have described the relationship of corticosteroid dose and AVN in systemic lupus erythematosus (SLE) and renal transplantation, although several have features limiting their generalizability. Many of the studies used a cross-sectional (case-control) methodology. An analysis of 22 studies of steroid-associated AVN determined there was a 4.6-fold increase in AVN occurrence for every 10 mg/day increase in oral steroid intake.[18] This study reported a strong correlation between mean daily dose and AVN risk but no correlation between peak dose and AVN. However, the study has significant limitations. Not all dose parameters were reported. The diagnosis relied on symptoms confirmed by X-ray; MRI was not used in the studies, thereby lowering the diagnostic sensitivity. Patients with SLE have an incidence of symptomatic AVN of 4–15% and up to 40% of asymptomatic AVN with typical AVN lesions seen on MRI.[18] A review of SLE patients treated with corticosteroids found the mean daily dose to be most correlative with AVN with 15–20% of patients developing AVN within 6–12 months of steroid treatment.[19–21] A contemporary review observed that studies done before 1978 and before the use of MRI often reported inaccurate associations between corticosteroid dose and AVN.[22] A compendium of case reports yielded the observation that patients with AVN had an average cumulative dose of 5,969 mg of prednisone. Approximately, 1 g of prednisone administered within a short time increased the risk of AVN.[22] Studies of SLE patients demonstrated associations between AVN and both daily dose and peak dose of corticosteroids. In a series of SLE patients with prednisone-associated AVN, the mean daily dose was greater than 20 mg/day, and the mean cumulative dose was 45 g, with a mean duration of treatment of 260 weeks.[23] One study found an association between AVN and corticosteroid doses >40 mg/day.[24] Another study reported a positive association at daily doses >30 mg for at least a month.[25] A third study reported a positive association at 60 mg for two months.[26] Several other studies reported associations with AVN variously with mean daily, peak, or cumulative dose.[27–29] Therefore, quantifiable dose thresholds are difficult to identify.

The situation may be more complicated in renal transplant patients with the use of cyclosporine and reductions in steroid doses. One study reported that AVN in this population is largely related to the daily steroid dose.[30] Associations between corticosteroid dose and AVN in renal transplant patients have generally concluded that a higher daily dose was associated with an increased risk of AVN.[18,31–33] Observations on cumulative dose are mixed because confounding factors, such as rejection episodes, have made studies heterogeneous. However, most recent studies do report an overall positive association between AVN and cumulative dose.[22]

We have studied two cohorts of patients, one with mixed etiologies and one with renal transplantation, all newly being treated with corticosteroids, with prospective screening MRIs of the hips for AVN.

The dose of corticosteroids was recorded and computed in prednisone equivalents as the peak, daily, and cumulative dose, and duration of exposure. Patients were studied with a limited screening coronal MRI before exposure to steroids and at six month intervals or sooner if hip pain occurred. A limited screening MRI was done with a protocol consisting of a coronal spin-ECHO T1-weighted sequence using the body coil (TR/TE 450–600/10–15, 6 mm slice thickness, 2 mm gap, 38 cm FOV, 256 × 256 matrix, 2 NEX) with the resultant acquisition of 12 images in three minutes and 30 seconds. A blinded comparison to a full hip MRI has demonstrated no statistically significant difference in the ability to detect femoral head AVN ($P = 0.48$). Compared to the full hip MRI, the diagnostic sensitivity, specificity, and accuracy of the limited screening MRIs were 98.0%, 99.2%, and 98.9%, respectively.[34]

Fifty patients newly exposed to corticosteroids for mixed etiologies other than renal transplantation were entered into the study. Forty-three completed the study, a lost to follow-up rate of 14%. Thirty-five percent of patients had inflammatory bowel disease, 35% had rheumatic diseases, 14% had respiratory diseases, and 16% had other inflammatory diagnoses. Three patients (7%) developed AVN detected by screening MRIs. These data are consistent with data reported previously where the incidence of AVN was 4–8%. No clinical or demographic features were identified as prevalent in patients with AVN but the number of patients who developed the disease was small. Corticosteroid doses in the two groups are presented in Table 1. Patients with AVN had higher cumulative doses than did patients in the same group without AVN ($P = 0.02$). Analysis of peak dose did not lend itself to similar statistical methods. No patient with AVN received <60 mg/day peak dose; 16/40 (40%) of patients without AVN

Table 1. Dose in patients with mixed etiologies (X ± SEM)

		Corticosteroid dose		
	n (%)	Duration (days)	Cumulative (mg)	Daily (mg)
AVN (−)	40 (93)	363.1 ± 57.4	4517.4 ± 762.7	15.7 ± 1.7
AVN (+)	3 (7)	710.3 ± 117.4	9496.8 ± 540.3	14.5 ± 3.6
p		0.10	0.02	0.72
Relative risk		5.8 at 700 days	8.8 at 9000 mg	1.3 at 20 mg

Table 2. Dose in transplant patients ($\overline{X} \pm$ SEM)

| | n (%) | Corticosteroid dose | | |
		Duration (days)	Cumulative (mg)	Daily (mg)
AVN (−)	27 (84)	816.1 ± 74.4	7576.1 ± 577.3	10.7 ± 0.8
AVN (+)	5 (16)	698.6 ± 80.0	8579.6 ± 476.6	13.4 ± 2.3
p		0.33	0.33	0.19
Relative risk		1.6 at 600 days	5.1 at 8000 mg	4.5 at 10 mg

received a peak dose of 60 mg/day, almost reaching statistical significance ($P = 0.08$). The relative risks of developing AVN for duration, cumulative dose, and daily dose are presented in the table. The relative risks for duration and cumulative dose were 5.8 and 8.8, respectively. All patients were diagnosed in stage 0 or I, prefracture stages in which hip preservation is possible.

Forty patients newly exposed to corticosteroids for kidney transplants were entered into the study. Thirty-two completed the study, a lost to follow-up rate of 20%. Five patients (16%) developed AVN. No differences in demographics or underlying diagnoses were found in patients with or without AVN. Corticosteroid exposure in this population is presented in Table 2. No correlations with corticosteroid doses were seen in transplant patients with and without AVN, suggesting that factors other than corticosteroid exposure accounted for the higher incidence of AVN in the transplant group. Again, the analysis of peak dose did not lend itself to the same statistical methods. In this cohort, all 27 patients without AVN also received a peak dose of 60 mg/day. The relative risks of cumulative and daily doses were 5.1 and 4.5, respectively. All patients were detected in stage 0, I. All patients in both cohorts were treated conservatively and no hip replacements were done. These data can be compared to consumption of ethanol of 400 mL/week that has been reported to present a relative risk of AVN of 3.3 and 400–1,000 mL/week where the relative risk rises to 9.8.[35]

It is an intriguing observation that the two clinical conditions most commonly seen in corticosteroid-associated AVN are SLE and renal transplantation. SLE adds the element of vasculitis whereas renal transplantation adds the decreased mineralization associated with renal osteodystrophy. The presence of antiphospholipid antibodies, producing a hyper-coagulable state in both SLE and renal transplantation, adds yet another potentially pathogenic mechanism. We, and others, have observed an increased frequency of elevated antiphospholipid antibodies in association with AVN in renal transplant patients. In our transplant database, the lupus anticoagulant was significantly elevated in patients with AVN compared to patients without AVN and carried a relative risk of 4.8. It was 93% specific in its association with AVN. Numerous studies have examined a correlation among anticardiolipin antibodies, coagulopathies, and corticosteroids in AVN with some, but not all, reporting associations.[28,36–38] One study of SLE patients with a 4.6% incidence of AVN found an association between AVN and antiphospholipid antibodies.[39] Another study reported that 54% of patients with SLE and AVN had antiphospholipid antibodies compared to 23% without AVN ($P < 0.05$).[21] These observations lead to the conclusion that AVN may require multiple "hits" or insults to the subchondral microcirculation, of which corticosteroids is one.[40–42]

Pathogenesis and histopathology of AVN

The etiology, natural history, and pathophysiology of AVN have recently been comprehensively reviewed, and salient features are briefly described here in the context of likely corticosteroid effects.[43] Broadly, the pathogenesis of AVN can be considered in two phases: (1) pathogenic events resulting in necrosis of marrow and osteocytes and, (2) a repair phase in which trabecular resorption exceeds bone formation leading to the structural compromise of subchondral trabeculae and subchondral fracture. Most studies are of the femoral head because it is the most frequent site of AVN but concepts of pathophysiology apply also to other subchondral bone sites.

Studies of the pathophysiologic events leading to cell death have focused on the vulnerable microcirculation in the femoral head and the consequences of vascular occlusion. A unifying concept of the pathogenesis of AVN has emphasized the central role of vascular occlusion and ischemia leading to osteocyte necrosis.[43] Decreased femoral head blood flow can occur through three mechanisms: vascular interruption by trauma (fracture or dislocation), extravascular compression (by lipocyte hypertrophy and marrow fat deposition), and thrombotic occlusion (by intravascular thrombi or embolic fat). The traumatic interruption of circulation to the femoral head in hip dislocations or displaced femoral neck fractures is the most obvious and well-understood pathogenic mechanism. The pathogenesis of non-traumatic AVN is considerably less well understood. Some investigators have suggested that both fibrin thrombi and embolic fat can occlude the microcirculation. Intravascular occlusion by fibrin thrombi or embolic fat has been found in a large number of specimens of AVN, and microcirculatory thrombosis associated specifically with fat emboli has been described.[44–48]

Regardless of the pathogenic mechanisms, ischemic necrosis of various etiologies eventually converges to a common and consistent structural compromise.[43] Osteocyte necrosis occurs after two to three hours of ischemia but histologic signs of osteocyte death are not apparent until 24–72 hours after the ischemic insult.[15,49–51] The earliest findings of AVN are necrosis of hematopoietic marrow and adipocytes and interstitial edema.[50,52] This is followed rapidly by osteocyte necrosis reflected by pyknosis of nuclei and empty osteocyte lacunae. Capillary neogenesis and revascularization occur to a degree in the necrotic zone.[53,54] With the entry of blood vessels into the zone of necrosis, a repair process begins consisting of coupled bone resorption and production that produces the radiographic appearance of sclerosis and lucency (Fig. 1).[51,53,55–57] Areas of lucency represent zones of bone resorption; areas of sclerosis are composed of both dead and living bone with the living reparative bone laminated onto dead trabeculae. The repair process is self-limited and incompletely replaces dead with living bone. In the subchondral region, bone formation occurs at a slower rate than does resorption, resulting in the net removal of bone, loss of structural integrity, subchondral fracture and collapse. It is not the necrosis *per se* but rather the repair process—particularly bone resorption—that leads to the loss of mechanical integrity of the femoral head, subchondral collapse, and joint incongruity. Finite element analysis has demonstrated the primary structural pathology to be resorption and weakening of subchondral trabeculae leading to subchondral fracture and joint incongruity.[58] The consequences are inevitable clinical and radiographic progression regardless of initial stage, requiring arthroplasty.[59,60]

Pathogenic effects of corticosteroids in AVN

The effects of corticosteroids on bone and bone vasculature are sufficiently complex to affect both the ischemic/necrotic and repair/resorption phases of AVN.[61] A number of studies have demonstrated that, after corticosteroid administration, lipid deposition in the extravascular marrow space and within osteocytes, together with adipocyte hypertrophy, produce an elevation in the intraosseous extravascular pressure and diminished blood flow similar to that of AVN of Gaucher's disease.[62] Hypertrophy and proliferation of adipocytes and abnormal lipid metabolism have been found in patients and experimental animals with AVN secondary to corticosteroid administration.[63–65] Corticosteroids can also elevate intraosseous pressure by increasing the synthesis of vasoactive peptides and increasing peripheral vascular resistance.

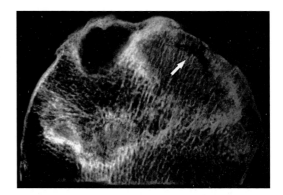

Figure 1. Specimen radiograph of AVN of the femoral head. Areas of sclerosis reflect necrotic and living bone, and lucent areas reflect bone resorption. Arrow indicates a fracture through resorbed subchondral trabeculae that predisposes to loss of sphericity and joint incongruity with pain and secondary osteoarthritis.

Corticosteroids also have direct effects on bone cells that can alter the repair/resorption phase of AVN and can contribute to the resorption of subchondral trabeculae and subchondral fractures. Corticosteroids dysregulate the balance of bone formation and resorption by inducing apoptosis of both osteoblasts and osteocytes, increasing the lifespan of osteoclasts, reducing the production of osteoblasts, and diminishing bone formation.[62,66,67] In addition to this mechanism, adipogenesis may also restrict the number of osteoprogenitor cells by shifting precursors from an osteocytic to an adipocytic lineage. Studies with a cloned mouse bone marrow progenitor cell line have indicated that adipocytes and osteocytes share a common progenitor cell, but in corticosteroid-induced or alcohol-induced adipogenesis, cells shunt from the osteocytic to the adipocytic lineage.[68] This is observed in cell-surface lineage markers and reflects the reduced ability of osteocytic cells to effect bone repair. Altering the homeostatic balance of bone turnover in favor of resorption at the expense of formation may contribute to bone loss in the mechanically sensitive subchondral trabeculae and subchondral fracture leading to joint incongruity and pain.

Early diagnosis of AVN

Recognizing that hip preservation is possible only in the prefracture stage, we undertook to prospectively screen with MRI the hips of patients with a variety of diseases including renal transplantation who had no previous exposure to corticosteroids. We compared *prospective* screening with MRI of patients considered to be at high risk of AVN with a *reactive* strategy of MRI and X-ray investigation of patients complaining of hip pain. Eighty patients newly started on corticosteroids were studied for two years with MRI every six months or sooner if hip pain occurred. The MRI and X-ray findings were compared to those in a population of 100 patients who presented with hip pain and who were subsequently diagnosed as having AVN. MRIs were graded with the University of Pennsylvania staging system in which stages 0–II are prefracture hips, stages III and IV have fractures of varying severity, and stages V and VI indicate joint space narrowing and arthritis.[69] Overall, nine (11%) of prospectively screened hips developed AVN. All were diagnosed in stage 0 or I, and all were able to be treated with hip sparing procedures, either bisphosphonates or core decompression. No

Table 3. Prospective and reactive screening with MRI

	U. Pennsylvania staging					
	0	I	II	III	IV	V
Prospective (AVN = 9/80)	3	6	0	0	0	0
Reactive (AVN = 100/100)	0	12	34	10	29	15

patient underwent hip replacement. By contrast, of 100 patients with hip pain diagnosed reactively as AVN, 12 were stage I, 34 were stage II, 10 stage III, 29 stage IV, and 15 stage V (Table 3). Forty-five percent of these patients underwent hip replacement within two years after diagnosis. These data confirm a study done of AVN after solid organ transplantation.[70] In that study, 103 hips of patients undergoing solid organ transplantation were studied prospectively with MRI. AVN was diagnosed in 8 of 103 hips (8%). All hips were diagnosed in Stage I. All cases of AVN of the femoral head developed within 10 months after the transplant. Seven of the eight hips with AVN were asymptomatic at the time of diagnosis. These data collectively suggest that identifying high-risk patients by understanding etiological relationships and subjecting at-risk populations to prospective screening can diagnose AVN at a prefracture stage where joint-preserving treatments can be used.

Conclusions

AVN is a progressive and disabling condition and, if untreated, leads to subchondral fracture and joint incongruity, usually requiring arthroplasty. Because joint preserving therapies are most effective before subchondral fracture, the key to successful treatment lies in identifying at-risk patients and diagnosing AVN before joint compromise. Surgical core decompression has been able to salvage approximately two-thirds of prefracture hips but is not useful in postfracture hips.[71,72] At the moment, joint preserving therapies are focused on the structural consequences of AVN and the prevention of trabecular resorption and subchondral fracture. Several studies have demonstrated that bisphosphonates can suppress osteoclastic resorption of subchondral trabeculae and prevent joint incongruity.[73] A better understanding of the pathogenesis of AVN might focus treatment on earlier stages of the disease to reverse or prevent mechanisms leading to ischemia. Corticosteroids have several deleterious effects on both the ischemic/necrotic and repair/resorption phases of AVN. Lovastatin has been shown to prevent corticosteroid-associated AVN by multifactorial mechanisms including counteracting the effects of corticosteroids on the differentiation of precursor cells in bone marrow from adipocytes to osteoblasts.[66] Because AVN can be detected by MRI before structural compromise, it would seem prudent to periodically screen patients newly exposed to corticosteroids and treat early lesions pharmacologically. This would be especially useful in patients undergoing renal transplantation or with SLE who are at particularly high risk of AVN.

Acknowledgments

This study was done with funding support from NIH Award 2K24AR002128–08. We gratefully acknowledge the Transplantation Service at Rhode Island Hospital, and Evan Cohen, M.D., David Kadmon, M.D., Sheldon Lidofsky, M.D., James Myers, M.D., Samir Shah, M.D., and Charles Sherman, M.D. for submitting patients to the study.

Conflict of interest

The authors declare no conflicts of interest.

References

1. Mankin, H.J. 1992. Nontraumatic necrosis of bone (osteonecrosis). *N. Engl. J. Med.* **326:** 1473–1479.
2. Zizic, T.M. 1985. Avascular necrosis of bone. In *Textbook of Rheumatology.* H.E. Kelly, W.N., S. Ruddy, *et al.*, Eds.: 1689–1710. WB Saunders. Philadelphia, PA.
3. Ono, K.S.Y. 1993. Epidemiology and risk factors in avascular necrosis of the femoral head. In *Bone Circulation and Vascularization in Normal and Pathologic Conditions.* A.J. Schoutens & J.W.M. Gardeniers, *et al.*, Eds.: 243–248. Plenum Press. New York, NY.
4. Boettcher, W.G. *et al.* 1970. Non-traumatic necrosis of the femoral head. I. Relation of altered hemostasis to etiology. *J. Bone Joint Surg. Am.* **52:** 312–321.
5. Hauzeur, J.P., J.L. Pasteels & S. Orloff. 1987. Bilateral nontraumatic aseptic osteonecrosis in the femoral head. An experimental study of incidence. *J. Bone Joint Surg. Am.* **69:** 1221–1225.
6. Jacobs, B. 1978. Epidemiology of traumatic and nontraumatic osteonecrosis. *Clin. Orthop. Relat. Res.* **130:** 51–67.
7. Marcus, N.D., W.F. Enneking & R.A. Massam. 1973. The silent hip in idiopathic aseptic necrosis. Treatment by bonegrafting. *J. Bone Joint Surg. Am.* **55:** 1351–1366.
8. Patterson, R.J., W.H. Bickel & D.C. Dahlin. 1964. Idiopathic avascular necrosis of the head of the femur. A study of fifty-two cases. *J. Bone Joint Surg. Am.* **46:** 267–282.
9. Bradway, J.K. & B.F. Morrey. 1993. The natural history of the silent hip in bilateral atraumatic osteonecrosis. *J. Arthroplasty* **8:** 383–387.
10. Vakil, N. & M. Sparberg. 1989. Steroid-related osteonecrosis in inflammatory bowel disease. *Gastroenterology* **96:** 62–67.
11. Cruess, R.L. 1981. Steroid-induced osteonecrosis. *J. R. Coll. Surg. Edinb.* **26:** 69–77.
12. Bradbury, G. *et al.* 1994. Avascular necrosis of bone after cardiac transplantation. Prevalence and relationship to administration and dosage of steroids. *J. Bone Joint Surg. Am.* **76:** 1385–1388.
13. Atsumi, T. & Y. Kuroki. 1992. Role of impairment of blood supply of the femoral head in the pathogenesis of idiopathic osteonecrosis. *Clin. Orthop. Relat. Res.* **277:** 22–30.
14. Barnes, R. *et al.* 1976. Subcapital fractures of the femur. A prospective review. *J. Bone Joint Surg. Br.* **58:** 2–24.
15. Bauer, T.A.S.B. 1993. The histology of osteonecrosis and its distinction from histologic artifacts. In *Bone Circulation and Vascularization in Normal and Pathological Conditions.* A.J. Schoutens & J.W.M. Gardeniers, *et al.*, Eds.: 283–292. Plenum. New York, NY.
16. Berger, C.E. *et al.* 2000. Elevated levels of lipoprotein(a) in familial bone marrow edema syndrome of the hip. *Clin. Orthop. Relat. Res.* **377:** 126–131.
17. Jones, J.P., Jr. 1994. Concepts of etiology and early pathogenesis of osteonecrosis. *Instr. Course Lect.* **43:** 499–512.
18. Felson, D.T. & J.J. Anderson. 1987. Across-study evaluation of association between steroid dose and bolus steroids and avascular necrosis of bone. *Lancet* **1:** 902–906.
19. Wolverton, S.E. 1995. Major adverse effects from systemic drugs: defining the risks. *Curr. Probl. Dermatol.* **7:** 6–38.
20. Weiner, E.S. & M. Abeles. 1989. Aseptic necrosis and glucocorticoids in systemic lupus erythematosus: a reevaluation. *J. Rheumatol.* **16:** 604–608.
21. Nagasawa, K. *et al.* 1989. Avascular necrosis of bone in systemic lupus erythematosus: possible role of haemostatic abnormalities. *Ann. Rheum. Dis.* **48:** 672–676.
22. Powell, C. *et al.* 2010. Steroid induced osteonecrosis: an analysis of steroid dosing risk. *Autoimmun. Rev.* **9:** 721–743.
23. Zizic, T.M. *et al.* 1985. Corticosteroid therapy associated with ischemic necrosis of bone in systemic lupus erythematosus. *Am. J. Med.* **79:** 596–604.
24. Massardo, L. *et al.* 1992. High-dose intravenous methylprednisolone therapy associated with osteonecrosis in patients with systemic lupus erythematosus. *Lupus* **1:** 401–405.
25. Ono, K., T. Tohjima & T. Komazawa. 1992. Risk factors of avascular necrosis of the femoral head in patients with systemic lupus erythematosus under high-dose corticosteroid therapy. *Clin. Orthop. Relat. Res.* **277:** 89–97.
26. Zonana-Nacach, A. *et al.* 2000. Damage in systemic lupus erythematosus and its association with corticosteroids. *Arthritis Rheum.* **43:** 1801–1808.
27. Oinuma, K. *et al.* 2001. Osteonecrosis in patients with systemic lupus erythematosus develops very early after starting high dose corticosteroid treatment. *Ann. Rheum. Dis.* **60:** 1145–1148.

28. Mont, M.A. *et al.* 1997. Risk factors for osteonecrosis in systemic lupus erythematosus. *J. Rheumatol.* **24:** 654–662.

29. Uea-areewongsa, P. *et al.* 2009. Factors associated with osteonecrosis in Thai lupus patients: a case control study. *J. Clin. Rheumatol.* **15:** 345–349.

30. Landmann, J. *et al.* 1987. Cyclosporin A and osteonecrosis of the femoral head. *J. Bone Joint Surg. Am.* **69:** 1226–1228.

31. Tang, S. *et al.* 2000. Risk factors for avascular bone necrosis after renal transplantation. *Transplant. Proc.* **32:** 1873–1875.

32. Lausten, G.S., J.S. Jensen & K. Olgaard. 1988. Necrosis of the femoral head after renal transplantation. *Acta Orthop. Scand.* **59:** 650–654.

33. Metselaar, H.J. *et al.* 1985. Incidence of osteonecrosis after renal transplantation. *Acta Orthop. Scand.* **56:** 413–415.

34. Khanna, A.J. *et al.* 2000. Femoral head osteonecrosis: detection and grading by using a rapid MR imaging protocol. *Radiology* 217: 188–192.

35. Matsuo, K. *et al.* 1988. Influence of alcohol intake, cigarette smoking, and occupational status on idiopathic osteonecrosis of the femoral head. *Clin. Orthop. Relat. Res.* **234:** 115–123.

36. Houssiau, F.A. *et al.* 1998. Magnetic resonance imaging-detected avascular osteonecrosis in systemic lupus erythematosus: lack of correlation with antiphospholipid antibodies. *Br. J. Rheumatol.* **37:** 448–453.

37. Mok, M.Y., V.T. Farewell & D.A. Isenberg. 2000. Risk factors for avascular necrosis of bone in patients with systemic lupus erythematosus: is there a role for antiphospholipid antibodies? *Ann. Rheum. Dis.* **59:** 462–467.

38. Gladman, D.D. *et al.* 2001. Predictive factors for symptomatic osteonecrosis in patients with systemic lupus erythematosus. *J. Rheumatol.* **28:** 761–765.

39. Asherson, R.A. *et al.* 1993. Avascular necrosis of bone and antiphospholipid antibodies in systemic lupus erythematosus. *J. Rheumatol.* **20:** 284–288.

40. Hirata, T. *et al.* 2007. Low molecular weight phenotype of Apo(a) is a risk factor of corticosteroid-induced osteonecrosis of the femoral head after renal transplant. *J. Rheumatol.* **34:** 516–522.

41. Aaron, R.K. & D.M. Ciombor. 2001. Coagulopathies and osteonecrosis. *Curr. Opin. Orthop.* **12:** 378–383.

42. Aaron, R.K., J. Sweeney, L. Bausserman, *et al.* 2011. *Coagulapothies in Avascular Necrosis of the Hip*. The Warren Alpert Medical School of Brown University. Providence, RI. In press.

43. Aaron, R.K. & R. Gray. 2007. Osteonecrosis: etiology, natural history, pathophysiology, and diagnosis. In *The Adult Hip*. Volume 1. J.J. Callaghan, A.G. Rosenberg, H.E. Rubash, Eds.: 465–476. Lippincott Williams & Wilkins. Philadelphia, PA.

44. Bonfiglio, M. 1976. Development of bone necrosis lesions. In *Underwater Physiology*. V.C. Lambertsen, Ed.: 117–132. Federation of American Societies for Experimental Biology. Bethesda, MD.

45. Ficat, R.a.A., J. 1980. Functional investigation of bone under normal conditions. In *Ischemia and Necrosis of Bone*. D.S. Hungerford, Ed.: 29–52. Williams & Wilkins. Baltimore, MD.

46. Jones, J.P., Jr. 1992. Intravascular coagulation and osteonecrosis. *Clin. Orthop. Relat. Res.* **277:** 41–53.

47. Saito, S., K. Ohzono & K. Ono. 1992. Early arteriopathy and postulated pathogenesis of osteonecrosis of the femoral head. The intracapital arterioles. *Clin. Orthop. Relat. Res.* **277:** 98–110.

48. Spencer, J.D. & M. Brookes. 1988. Avascular necrosis and the blood supply of the femoral head. *Clin. Orthop. Relat. Res.* **235:** 127–140.

49. James, J. & G.L. Steijn-Myagkaya. 1986. Death of osteocytes. Electron microscopy after in vitro ischaemia. *J. Bone Joint Surg. Br.* **68:** 620–624.

50. Jones, J.P., Jr., 1985. Osteonecrosis. In *Arthritis and Allied Conditions*. D. McCarty, Ed.: 1356–1373. Lea & Febiger. Philadelphia, PA.

51. Kenzora, J.E., M.J. Glimcher. 1981. Osteonecrosis. In *Textbook of Rheumatology*, H.E. Kelly W.N., S. Ruddy *et al.*, Eds.: 1755–1782. WB Saunders. Philadelphia, PA.

52. Saito, S., A. Inoue & K. Ono. 1987. Intramedullary haemorrhage as a possible cause of avascular necrosis of the femoral head. The histology of 16 femoral heads at the silent stage. *J. Bone Joint Surg. Br.* **69:** 346–351.

53. Glimcher, M.J. & J.E. Kenzora. 1979. Nicolas Andry award. The biology of osteonecrosis of the human femoral head and its clinical implications: 1. Tissue biology. *Clin. Orthop. Relat. Res.* **138:** 284–309.

54. Jones, J.P., Jr. 1991. Etiology and pathogenesis of osteonecrosis. *Semin. Arthroplasty* 2: 160–168.

55. Catto, M. 1965. A histological study of avascular necrosis of the femoral head after transcervical fracture. *J. Bone Joint Surg. Br.* **47:** 749–776.

56. Glimcher, M.J. & J.E. Kenzora. 1979. The biology of osteonecrosis of the human femoral head and its clinical implications: II. The pathological changes in the femoral head as an organ and in the hip joint. *Clin. Orthop. Relat. Res.* **139:** 283–312.

57. Glimcher, M.J. & J.E. Kenzora. 1979. The biology of osteonecrosis of the human femoral head and its clinical implications. III. Discussion of the etiology and genesis of the pathological sequelae; comments on treatment. *Clin. Orthop. Relat. Res.* **140:** 273–312.

58. Brown, T.D., K.J. Baker & R.A. Brand. 1992. Structural consequences of subchondral bone involvement in segmental osteonecrosis of the femoral head. *J. Orthop. Res.* **10:** 79–87.

59. Aaron, R.K., B.N. Stulberg & D.W. Lennox. 2001. Clinical and radiographic outcomes in untreated symptomatic osteonecrosis of the femoral head. *Tech. Orthop.* **16:** 1–5.

60. Aaron, R.K., B.N. Stulberg & D.W. Lennox. 1997. The natural history of osteonecrosis of the femoral head and risk factors for progression. In *The Etiology, Diagnosis and Management of Osteonecrosis of the Human Skeleton*. J.R. Urbaniak, Ed.: 261–265. American Academy of Orthopaedic Surgeons. Chicago, IL.

61. Kerachian, M.A., C. Séguin & E.J. Harvey. 2009. Glucocorticoids in osteonecrosis of the femoral head: a new understanding of the mechanisms of action. *J. Steroid Biochem. Mol. Biol.* **114:** 121–128.

62. Wang, G.J., Q. Cui & G. Balian. 2000. The Nicolas Andry award. The pathogenesis and prevention of steroid-induced osteonecrosis. *Clin. Orthop. Relat. Res.* **370:** 295–310.

63. Kawai, K., A. Tamaki & K. Hirohata. 1985. Steroid-induced accumulation of lipid in the osteocytes of the rabbit femoral

head. A histochemical and electron microscopic study. *J. Bone Joint Surg. Am.* **67:** 755–763.

64. Wang, G.J. *et al.* 1981. Cortisone-induced intrafemoral head pressure change and its response to a drilling decompression method. *Clin. Orthop. Relat. Res.* **159:** 274–278.

65. Wang, G.J. *et al.* 1977. Fat-cell changes as a mechanism of avascular necrosis of the femoral head in cortisone-treated rabbits. *J. Bone Joint Surg. Am.* **59:** 729–735.

66. Cui, Q., G.J. Wang & G. Balian. 1997. Steroid-induced adipogenesis in a pluripotential cell line from bone marrow. *J. Bone Joint Surg. Am.* **79:** 1054–1063.

67. Wang, G.J. & Q. Cui. 1997. The pathogenesis of steroid-induced osteonecrosis and the effect of lipid-clearing agents on this mechanism. In *The Pathogenesis of Steroid-Induced Osteonecrosis and the Effect of Lipid-Clearing Agents on this Mechanism*. J.R. Urbaniak, & J.P. Jones, Eds.: 159–166. American Academy of Orthopaedic Surgeons. Rosemont, IL.

68. Wang, Y. *et al.* 2003. Alcohol-induced adipogenesis in bone and marrow: a possible mechanism for osteonecrosis. *Clin. Orthop. Relat. Res.* **410:** 213–224.

69. Steinberg, M.E., G.D. Hayken & D.R. Steinberg. 1995. A quantitative system for staging avascular necrosis. *J. Bone Joint Surg. Br.* **77:** 34–41.

70. Marston, S.B. *et al.* 2002. Osteonecrosis of the femoral head after solid organ transplantation: a prospective study. *J. Bone Joint Surg. Am.* **84-A:** 2145–2151.

71. Aaron, R.K. *et al.* 1989. The conservative treatment of osteonecrosis of the femoral head. A comparison of core decompression and pulsing electromagnetic fields. *Clin. Orthop. Relat. Res.* **249:** 209–218.

72. Mont, M.A. & D.S. Hungerford. 1995. Non-traumatic avascular necrosis of the femoral head. *J. Bone Joint Surg. Am.* **77:** 459–474.

73. Agarwala, S. *et al.* 2005. Efficacy of alendronate, a bisphosphonate, in the treatment of AVN of the hip. A prospective open-label study. *Rheumatology (Oxford)* **44:** 352–359.

Ann. N.Y. Acad. Sci. ISSN 0077-8923

ANNALS OF THE NEW YORK ACADEMY OF SCIENCES

Issue: *Skeletal Biology and Medicine II*

Measuring intranodal pressure and lymph viscosity to elucidate mechanisms of arthritic flare and therapeutic outcomes

Echoe M. Bouta,[1,2] Ronald W. Wood,[3,4] Seth W. Perry,[2] Edward B. Brown,[2] Christopher T. Ritchlin,[1,5] Lianping Xing,[1,6] and Edward M. Schwarz[1,2,4,5,6]

[1]Center for Musculoskeletal Research, [2]Department of Biomedical Engineering, [3]Department of Obstetrics and Gynecology, [4]Department of Urology, [5]Division of Allergy, Immunology, and Rheumatology, and [6]Department of Pathology and Laboratory Medicine, University of Rochester School of Medicine and Dentistry, Rochester, New York

Address for correspondence: Dr. Edward M. Schwarz, The Center for Musculoskeletal Research, University of Rochester Medical Center, 601 Elmwood Avenue, Box 665, Rochester, NY 14642. Edward_Schwarz@URMC.Rochester.edu

Rheumatoid arthritis (RA) is a chronic autoimmune disease with episodic flares in affected joints; the etiology of RA is largely unknown. Recent studies in mice demonstrated that alterations in lymphatics from affected joints precede flares. Thus, we aimed to develop novel methods for measuring lymph node pressure and lymph viscosity in limbs of mice. Pressure measurements were performed by inserting a glass micropipette connected to a pressure transducer into popliteal lymph nodes (PLN) or axillary lymph nodes (ALN) of mice; subsequently, we determined that the lymphatic pressures of water were 9 and 12 cm, respectively. We are also developing methods for measuring lymph viscosity in lymphatic vessels afferent to PLN, which can be measured by multiphoton fluorescence recovery after photobleaching (MP-FRAP) of fluorescein isothiocyanate–labeled bovine serum albumin (FITC-BSA) injected into the hind footpad. These results demonstrate the potential of lymph node pressure and lymph viscosity measurements, and future studies to test these outcomes as biomarkers of arthritic flare are warranted.

Keywords: rheumatoid arthritis; lymph node; flare; lymphatic pressure; lymph viscosity

Introduction

Rheumatoid arthritis (RA) is a debilitating immune-mediated inflammatory disorder characterized by joint inflammation and destruction, which affects 1% of the population.[1] Major advances in our understanding of pivotal events that underlie the pathobiology of RA have fostered the development of several biologic therapies that specifically target the inflammatory cytokines (TNF-α, IL-1, IL-6) and immune cells (B cells, monocytes) implicated in synovitis and joint destruction. Despite these advances, several aspects of RA pathogenesis remain poorly understood. Foremost among them are the episodic flares and remissions observed in RA over the course of time. Joint flares are associated with significant morbidity and loss of function, so a major unmet need is the elucidation of flare mech-

anisms to catalyze new drug development for RA patients.

Murine models of acute and chronic RA, such as the K/BxN serum-induced arthritis (SIA)[2] and TNF transgenic mouse (TNF-Tg)[3] models, respectively, have been very useful for elucidating the pathophysiology of inflammatory-erosive arthritis. Recently, inflamed joints in these models have been analyzed with contrast enhanced magnetic resonance imaging (CE–MRI). These imaging studies demonstrate that arthritic knee flare is associated with the expansion and subsequent collapse of the popliteal lymph node (PLN),[4–7] which drains inflamed joints.[7] This PLN collapse, that is, reduction in lymph node volume, is accompanied by the migration of a subset of $CD21^{hi}CD23^{+}IgM^{hi}CD1d^{+}$ B cells in inflamed nodes (Bin) into the paracortical sinuses and this centripetal movement is associated with blockage

doi: 10.1111/j.1749-6632.2011.06237.x

of the lymphatic vasculature: a process that is ameliorated by anti-CD20 B cell depletion therapy.[6] Additional alterations in lymphoid biology during inflammatory arthritis include increased VEGF-C dependent lymphangiogenesis[8,9] and diminished lymphatic pulsation as determined by near infrared-indocyanine green (NIR-ICG) imaging.[10] Thus, a formal association between altered lymphatic volumes and variation in lymphatic flow and the onset of arthritic flare has been established. Unfortunately, methods to directly measure the lymphatic pressure and viscosity changes during and after arthritic flare and following therapy have not been established in animal models of RA.

To facilitate a better understanding of lymph node pressure and lymph viscosity, we explored several *in vivo* imaging techniques in normal wild-type (WT) mice. The results demonstrated that pressure can be measured using a glass micropipette connected to a pressure transducer, while viscosity measurements are feasible by using the Stokes–Einstein equation and calculating the diffusion coefficient following *in vivo* MP-FRAP.

Materials and methods

Animals
The female WT C57Bl/6 mice used in this study were purchased from Charles River Laboratories (Wilmington, MA). All mice were between 8 and 10 weeks old. During NIR-ICG imaging, animals were anesthetized with 1.5% isoflurane in oxygen. During MP-FRAP, animals were anesthetized with ketamine. The research was conducted with approval by the University of Rochester Medical Center Institutional Animal Care and Use Committee.

Pressure measurements
Mice were injected with 30 μL of 1% Evan's blue (Sigma Aldrich, St. Louis, MO) in the hind footpad or front paw for initial visualization of the PLN ($n = 8$) or ALN ($n = 7$), respectively. After approximately 20 min, the PLN or ALN was exposed via an incision through the overlying skin. An operating microscope (Op Mi6, Ziess, West Germany) was used to insert a luer lock borosilicate glass micropipette with an inner tip diameter of 30 μm (World Precision Instruments, Sarasota, FL) into the PLN or ALN. The glass micropipette was filled with ICG and attached to a pressure transducer (Sorenson Transpac IV, Abbott,

North Chicago, IL) and infusion pump (PHD 2000, Harvard Apparatus, Holliston, MA) via polyvinyl chloride (PVC) tubing (Fig. 1A). Correct placement was confirmed visually and by pumping ICG into the PLN through the micropipette and confirming that it resided in the PLN and upstream lymphatics (Fig. 1B). Pressure recording and control of ICG imaging parameters was achieved using customized software (LabVIEW; National Instruments, Austin, TX). If excessive leaking of ICG was detected (Fig. 1C), indicating improper placement of the micropipette, data were discarded. Calibration of the glass micropipette and pressure transducer was preformed after each session manometrically.

Viscosity measurements by MP-FRAP
While FRAP has been used in biological systems to characterize flow,[11,12] MP-FRAP has the

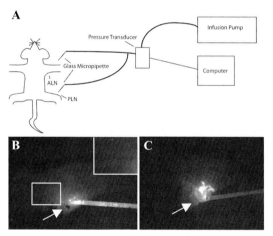

Figure 1. Schematic diagram of a mouse instrumented with the PLN and ALN pressure measurement system and NIR imaging to detect leakage. Mice are anesthetized with isoflurane, placed on a heated surgery table, and Evan's blue dye is injected into the hind footpad or front paw, respectively. Twenty minutes later, a surgical incision is made to expose either the PLN or ALN, and the ICG-filled borosilicate glass micropipette connected to the pressure transducer and infusion pump is placed into the lymph node. The pressure measurement is then obtained without flow from the pump. To confirm proper placement of the micropipette, the ICG flow into the lymph node during increasing pressure from the infusion pump is monitored by NIR imaging. A schematic representation of this procedure is presented (A). Only pressure measurements from lymph nodes that have not been ruptured by the glass micropipette are recorded. This is evident in nonleaking nodes (approximately 80% of injections) by a limited residue ICG signal at the injection site (solid arrow in B), versus the gross ICG leakage of a ruptured node (solid arrow in C). ICG can be seen upstream in lymph nodes that have been successfully injected with ICG (inset in B).

advantage of being able to be performed *in vivo* in real time.[13] Briefly, a laser is used to monitor the sample at a low power while keeping focal volume consistent throughout the experiment. Then, the sample is photobleached by a brief strong pulse of the laser, modulated through a Pockels' cell, to induce photobleaching. After the bleach, the laser power returns to the lower monitor power as the fluorescence recovers to baseline. Photons are collected by a photomultiplier tube and recorded throughout the experiment by a photon counter. Multiple monitor–bleach–recovery curves, performed in rapid succession, are summed to produce the final curve. Recovery is dependent on diffusion and convection. From the recovery curve after photobleaching, the diffusion coefficient and the flow rate can be determined. The diffusion coefficient is found by fitting the recovery curve to the following equation in Matlab (MathWorks, Natick, MA):[13,14]

$$F(t) = F_0 \sum_{n=0}^{\infty} \frac{(-\beta)^n}{n!}$$
$$\times \frac{\exp\left[\dfrac{-4n\left(t/\tau_v^2\right)}{1 + n + 2nt/\tau_D}\right]}{(1 + n + 2nt/\tau_D)\left(1 + n + 2nt/R\tau_D\right)^{1/2}} \quad (1)$$

where β is the bleach depth parameter, τ_v is a time constant due to flow, τ_D is a time constant due to diffusion, and R is the square of the ratio of the axial to radial dimensions of the focal volume.

From this, the diffusion coefficient (D) can be determined by[13,14]

$$D = w_r^2 / 8\tau_D \quad (2)$$

where w_r is the radius of the focal volume.

Viscosity, η, can be found from the diffusion coefficient by the Stokes–Einstein equation:

$$\eta = \frac{k_b T}{6\pi D r} \quad (3)$$

where k_b is Boltzmann's constant, T is temperature and r is the hydrodynamic radius of FITC-BSA.

In our *in vivo* paradigm, MP-FRAP is first calibrated by measuring the *in vitro* diffusion coefficient of FITC-BSA in water. Because the viscosity of water is a constant, and FITC-BSA D values can be fit by models,[13,14] the relative FITC-BSA D values calculated *in vitro* (in water) versus *in vivo* (in lymph) can be used to ratiometrically calculate the viscosity of lymph via the Stokes–Einstein equation

(Eq. 3). This ratiometric approach may help minimize sources of experimental uncertainty in the final calculated lymph viscosity, compared to calculating viscosity from the Stokes–Einstein equation alone based on a software-fitted D value for FITC-BSA in lymph (which, in turn, is dependent on lymph viscosity). The hind footpads of mice were injected with 30 μL of FITC-BSA (Sigma Aldrich, St. Louis, MO) ≤2 h before MP-FRAP measurements. Correct location of MP-FRAP was confirmed by imaging the lymphatic vessel via two-photon microscopy. Only curves with a smooth recovery and sufficient bleaching were used.

Statistical analysis

All results are presented as the mean ± standard deviation (SD). Data were first checked for normality and groups were checked to have equal variances by the KS test and *F*-test, respectively. Comparisons between groups were analyzed by a two-tailed Student's *t*-test. *P* values less than 0.05 were considered significant.

Results

The lymphatic pressure was measured in both popliteal and axillary lymph nodes of WT mice using the system described in Figure 1. The pressure of axillary lymph nodes was 11.70 ± 0.83 cm H_2O and was 25% greater than that observed in PLN (8.95 ± 0.73 cm H_2O) (Fig. 2A).

Occasionally, a pulse could be detected in the pressure measurement (Fig. 2B). This pulse rate (1 beat/30 sec) is consistent with the known intrinsic lymphatic pulse that we observed using NIR imaging in mouse legs.[10] Although we did succeed in measuring the lymphatic pulse rate with our method, this approach will need to be markedly improved to detect it consistently, as it likely requires precise positioning of the glass micropipette within the lymphatic channels running through the lymph node. This can likely be achieved with a precise micromanipulator.

As described in the Materials and methods section, the viscosity of lymph in lymphatic vessels afferent to PLN is determined *in vivo* by first measuring the diffusion coefficient of known standards *in vitro* (Fig. 3A). Others have calculated the *in vitro* diffusion coefficient of FITC-BSA in water to be ~ 50 μm^2/second.[13,14] *In vivo* MP-FRAP curves (Fig. 3B) are then acquired from perivalve regions of

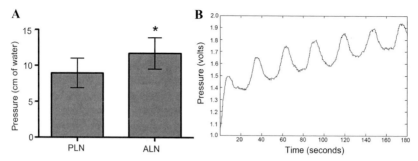

Figure 2. *In vivo* axillary and popliteal lymph node pressure measurements. *In vivo* pressure measurements of axillary and popliteal lymph nodes of mice ($n = 8$ for PLN, $n = 7$ for ALN) were obtained using the methods described in Figure 1. (A) The data are presented as the mean $\pm$ SD ($*P < 0.05$ vs. PLN). Occasionally, a pulse in the pressure measurement could be detected, indicating direct placement of the micropipette into a lymphatic channel, as evidenced by a sinusoidal curve of the rhythmic lymphatic pulse (B).

lymphatic vessels that have been filled with FITC-BSA by injecting the hind footpad (Fig. 3C). We are presently working on adapting our computer modeling to obtain best fits of fluorescent recovery D values for the *in vivo* lymphatic environment, but these results demonstrate that acquisition of MP-FRAP data from lymph vessels *in vivo* are feasible and should ultimately yield real-time values of lymph viscosity.

Discussion

Here, we describe novel methods to quantify pressure in draining lymph nodes and viscosity in lymphatic vessels and PLN of mice, and propose that they will be helpful toward elucidating the mechanisms during and after arthritic flare and responses to therapy using various murine models of RA.[15] We hypothesize that both pressure of the lymph

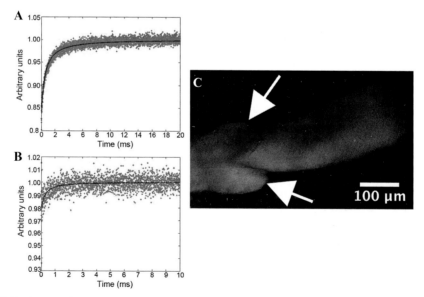

Figure 3. MP-FRAP approach to measure *in vivo* viscosity of lymph in lymphatic vessels afferent to PLN. First, MP-FRAP plots of a 1mg/mL *in vitro* solution of FITC-BSA–are acquired (normalized fluorescence photon counts, F_T/F_0) (A). Then *in vivo* MP-FRAP plots are acquired from FITC-BSA–loaded lymphatic vessels afferent to PLN. For the *in vivo* MP-FRAP, 30 μL of 1mg/mL FITC-BSA was injected into the hind footpad, and lymphatic vessels were surgically exposed $\leq$2 h later, followed by MP-FRAP (B). A two-photon image of the FITC-BSA–loaded lymphatic vessel, from which MP-FRAP data were obtained, is shown (arrows indicate the valves in the lymphatic vessel) (C).

nodes and viscosity of lymph will increase during the pathogenesis of inflammatory arthritis. This pressure increase could be caused by the influx of macrophages and Bin cells into the draining lymph node,[6] and the fivefold increase in lymphatic pulsing.[10] Additional lymphatic pressure could be caused by a change in viscosity due to bone and joint catabolism in which extracellular matrix breakdown products and minerals are cleared in lymph. While lymphangiogenesis and draining lymph node expansion are designed to counteract this pressure increase and prevent lymphatic vessel rupture, there is likely a threshold pressure that exceeds the capacity of these compensatory mechanisms. We hypothesize that when this threshold pressure is achieved, it triggers shutdown of the lymphatic pulse, which results in the collapse of the draining lymph node. The resulting loss of lymphatic draining from the joint is manifested as an arthritic flare.

Despite its known importance as a biomarker of inflammation and tumor metastasis, very little has been done to measure the parameters of the murine lymphatic system because of the diminutive size of the animal compared to other species used for lymphatic investigations. Pressure has been measured in the lymphatic vessels and capillaries in humans, sheep, and rabbits.[16–18] The measurement of lymphatic pressure in small mammals is scarce, but an early study measured pressure of lymphatic vessels in mouse ear.[19] However, there is no report of measuring lymphatic pressure in draining lymph nodes of mouse limbs where arthritis occurs. Recently, others have measured the pressure inside the human lymph node for detection of cancer cell metastasis in the lymph nodes. The study found that the intranodal pressure in sentinel lymph nodes without tumors was 9.1 ± 6.2 mm Hg,[20] very similar to our findings. Tumor-containing lymph nodes were found to have a pressure of 21.4 ± 15.4 mm Hg; the rationale is that when more cells reside within the lymphatic vessels of a sentinel lymph node, it will increase overall pressure of the node.

Although we have no formal explanation for the difference in pressure seen between the ALN and PLN, one possible contributor is the remarkable difference in mass between the lymphatics of the lower versus upper limbs of mice. As the volume of the lymphatics of the lower limb is apparently greater than the upper limb, we hypothesize that distribu-

tion over a greater surface area could result in lower pressure.

Others have measured the viscosity of lymph in dogs by collecting the lymph from the thoracic duct and using a viscometer. The viscosity of lymph was analyzed to determine if changes of lymph correlate with changes in diet.[21] Unfortunately, parallel approaches are not applicable in smaller animals where the lymphatic vessels cannot be cannulated, and low lymph volumes preclude application of typical methods to determine viscosity. MP-FRAP overcomes these constraints by using optical techniques, opposed to physically measuring the viscosity of the fluid in question. This allows the viscosity of the fluid to be found without being collected so the viscosity of lymph can be measured throughout the course of the experiment.

The methods described in this paper overcome the size constraints of measuring murine lymph node pressure and lymph viscosity. Pressure can be measured by inserting a glass micropipette into the PLN while the viscosity of lymph can be measured by MP-FRAP. Experiments designed to test our hypotheses that both lymph node pressure and lymph viscosity will increase during the pathogenesis of inflammatory arthritis are currently underway using the methods described here.

Acknowledgments

This work was supported by research grants from the National Institutes of Health PHS Awards (T32 AR053459; R01s AR048697, AR053586, and AR056702; P01 AI078907; DP2OD006501; and P30 AR061307).

Conflicts of interest

Dr. Ronald Wood is an inventor of lymphatic imaging techniques that the University of Rochester has licensed to Novadaq Technologies, Inc.

References

1. Firestein, G.S. 2003. Evolving concepts of rheumatoid arthritis. *Nature* **423:** 356–361.
2. Kouskoff, V., A.S. Korganow, V. Duchatelle, *et al.* 1996. Organ-specific disease provoked by systemic autoimmunity. *Cell. Mol. Bioeng.* **87:** 811–822.
3. Keffer, J., L. Probert, H. Cazlaris, *et al.* 1991. Transgenic mice expressing human tumour necrosis factor: a predictive genetic model of arthritis. *EMBO* **10:** 4025–4031.

4. Proulx, S.T., E. Kwok, Z. You, *et al.* 2007. MRI and quantification of draining lymph node function in inflammatory arthritis. *Ann. N. Y. Acad. Sci.* **1117:** 106–123.

5. Proulx, S.T., E. Kwok, Z. You, *et al.* Longitudinal assessment of synovial, lymph node, and bone volumes in inflammatory arthritis in mice by in vivo magnetic resonance imaging and microfocal computed tomography. *Arthritis Rheum.* **56:** 4024–4037.

6. Li, J., I. Kuzin, S. Moshkani, *et al.* 2010. Expanded CD23(+)/CD21(hi) B cells in inflamed lymph nodes are associated with the onset of inflammatory-erosive arthritis in TNF-transgenic mice and are targets of anti-CD20 therapy. *J. Immunol.* **184:** 6142–6150.

7. Guo, R., Q. Zhou, S.T. Proulx, *et al.* 2009. Inhibition of lymphangiogenesis and lymphatic drainage via vascular endothelial growth factor receptor 3 blockade increases the severity of inflammation in a mouse model of chronic inflammatory arthritis. *Arthritis Rheum.* **60:** 2666–2676.

8. Zhang, Q., Y. Lu, S.T. Proulx, *et al.* 2007. Increased lymphangiogenesis in joints of mice with inflammatory arthritis. *Arthritis Res. Ther.* **9:** R118.

9. Zhou, Q., R. Guo, R. Wood, *et al.* 2011. Vascular endothelial growth factor C attenuates joint damage in chronic inflammatory arthritis by accelerating local lymphatic drainage in mice. *Arthritis Rheum.* **63:** 2318–2328.

10. Zhou, Q., R. Wood, E.M. Schwarz, *et al.* 2010. Near-infrared lymphatic imaging demonstrates the dynamics of lymph flow and lymphangiogenesis during the acute versus chronic phases of arthritis in mice. *Arthritis Rheum.* **62:** 1881–1889.

11. Kwon, R.Y. & J.A. Frangos. 2010. Quantification of lacunar-canalicular interstitial fluid flow through computational modeling of fluorescence recovery after photobleaching. *Cell Mol. Bioeng.* **3:** 296–306.

12. Bonvin, C., J. Overney, A.C. Shieh, *et al.* 2010. A multichamber fluidic device for 3D cultures under interstitial flow with live imaging: development, characterization, and applications. *Biotechnol. Bioeng.* **105:** 982–991.

13. Sullivan, K.D., W.H. Sipprell 3rd, E.B. Brown Jr. & E.B. Brown 3rd. 2009. Improved model of fluorescence recovery expands the application of multiphoton fluorescence recovery after photobleaching in vivo. *Biophys. J.* **96:** 5082–5094.

14. Brown, E.B., *et al.* 1999. Measurement of molecular diffusion in solution by multiphoton fluorescence photobleaching recovery. *Biophys. J.* **77:** 2837–2849.

15. van den Berg, W.B. 2005. Animal models of arthritis. What have we learned? *J. Rheumatol. Suppl.* **72:** 7–9.

16. Franzeck, U.K., M. Fischer, U. Costanzo, *et al.* 1996. Effect of postural changes on human lymphatic capillary pressure of the skin. *J. Physiol.* **494:** 595–600.

17. McGeown, J.G., N.G. McHale & K.D. Thornbury. 1987. The effect of electrical stimulation of the sympathetic chain on peripheral lymph flow in the anaesthetized sheep. *J. Physiol.* **393:** 123–133.

18. Negrini, D. & M. Del Fabbro. 1999. Subatmospheric pressure in the rabbit pleural lymphatic network. *J. Physiol.* **520:** 761–769.

19. McMaster, P.D. 1947. The relative pressures within cutaneous lymphatic capillaries and the tissues. *J. Exp. Med.* **86:** 293–308.

20. Nathanson, S.D. & M. Mahan. 2011. Sentinel lymph node pressure in breast cancer. *Ann. Surg. Oncol.* doi: 10.1245/s10434-011-1796-y. [Epub ahead of print].

21. Burton-Optiz, R. & R. Nemser. 1917. The viscosity of lymph. *Am. J. Physiol.* **45:** 25–29.

Ann. N.Y. Acad. Sci. ISSN 0077-8923

ANNALS OF THE NEW YORK ACADEMY OF SCIENCES

Issue: *Skeletal Biology and Medicine II*

Genetic determinants of Paget's disease of bone

Stuart H. Ralston and Omar M. E. Albagha

Rheumatic Diseases Unit, Molecular Medicine Centre, University of Edinburgh, United Kingdom

Address for correspondence: Professor Stuart H. Ralston, Molecular Medicine Centre, Western General Hospital, Edinburgh EH4 2XU, UK. stuart.ralston@ed.ac.uk

Paget's disease is a common skeletal disorder with a strong genetic component, which is characterized by focal increases in disorganized bone remodeling, predominantly affecting the axial skeleton. Current evidence suggests that classical Paget's disease of bone (PDB) is caused by a combination of rare alleles of large effect size that cause autosomal dominant inheritance of the disease and more common alleles of smaller effect size. Mutations of *SQSTM1* are the most common cause of classical PDB, occurring in about 10% of patients. The causal mutations cluster in the ubiquitin-associated domain and impair its ability to bind ubiquitin. Other loci that predispose to PDB have recently been identified by genome-wide association studies, which have identified variants at seven loci that predispose to the disease. These increase the risk of PDB individually by 1.3- to 1.7-fold, but have combined effects that account for about 86% of the population-attributable risk of PDB in *SQSTM1* negative patients.

Keywords: genetic; genome-wide association study; Paget's disease; SNP

Epidemiology

Paget's disease of bone (PDB) affects between 1% and 2% of Caucasians over the age of 55 years. The disease is characterized by focal abnormalities of increased bone turnover affecting one or more sites throughout the skeleton. The axial skeleton is preferentially affected and the most common sites of involvement are the pelvis (70% of cases), femur (55%), lumbar spine (53%), skull (42%), and tibia (32%).[1] The prevalence of Paget's disease increases with age, and the disease affects about 8% of men and 5% of women by the eighth decade in the UK.[2] The UK has the highest incidence of PDB in the world, but the disease is also common in Western and Southern Europe[3] and in British migrants to Australia, New Zealand, and South Africa.[4] Conversely, PDB is rare in Scandinavia, the Indian subcontinent, China, Japan, and other countries in the Far East.[5,6] These ethnic differences in the incidence of PDB emphasize the importance of genetic factors in the pathogenesis of the disease, but there is evidence that environmental factors also play a role. Support for this comes from the observation that there have been reductions in the prevalence and severity of PDB in the UK, some European countries,[7] and New Zealand over the past 25 years,[8,9] and a delayed onset of the disease in children of patients with *SQSTM1*-mediated familial PDB.[10] However, for reasons that are currently unclear, no major changes in severity or incidence of PDB have been observed in other countries such as Italy[11] and some regions of the United States.[12] It is likely that there have been changes in hitherto unidentified environmental triggers for the disease in several countries over the past 25 years. However, the reduction in disease prevalence in some countries, such as the UK, may also be due in part to changes in ethnic makeup of the population due to influx of migrants from low prevalence regions such as the Indian subcontinent and the Far East.

Genetic architecture of PDB

Familial clustering is common in classical PDB and in many cases the disease is inherited as an autosomal dominant trait with a penetrance of about 80% and 90% by the seventh decade.[13–15] Approximately 15% of PDB cases report having a positive family history of the condition but the proportion of familial cases is likely to be higher because PDB is

doi: 10.1111/j.1749-6632.2011.06228.x

often asymptomatic.[16,17] The risk of developing PDB in relatives of an affected person is approximately seven times greater than in relatives of controls.[16–18] In addition, several rare syndromes have been described with a Mendelian mode of inheritance (either autosomal dominant or autosomal recessive) that share several features in common with PDB. It seems likely that the ethnic differences in incidence of PDB referred previously is due to differences between population in carriage of risk alleles for the disease, but this has not yet been specifically studied. These genes and loci that predispose to PDB are discussed in more detail later.

Genes that cause rare PDB-like syndromes

Several rare bone disorders have been described which exhibit clinical, radiological, or histological features in common with Paget's disease. These include: familial expansile osteolysis (FEO),[19] expansile skeletal hyperphosphatasia (ESH),[20] early onset familial Paget's disease (EoPDB),[21] juvenile Paget's disease (also known as idiopathic hyperphosphatasia),[22] and the syndrome of hereditary inclusion body myopathy, Paget disease of bone, and frontotemporal dementia (IBMPFD).[23] With the exception of juvenile Paget's disease, which is an autosomal recessive condition, all of these disorders are inherited in an autosomal dominant manner.

Mutations of the *TNFRSF11A* gene, which encodes RANK are responsible for the syndromes of FEO, ESH, and EoPDB. These disorders share several features in common including the presence of focal osteolytic lesions, premature deafness, and premature tooth loss.[21,22,24] The causal mutations are duplications of between 15–27 nucleotides in the first exon of *TNFRSF11A*, which introduce between 5 and 9 additional amino acids residues into the RANK signal peptide and prevent it from being cleaved normally.[24] This causes the abnormal RANK molecules to accumulate in the Golgi apparatus.[25] Cells which have been engineered to express these mutations do not show evidence of a constitutive increase in NF-κB signaling and also fail to activate NF-κB signaling in response to RANKL.[25] Despite this, overexpression of the mutants in osteoclast precursors promotes osteoclast differentiation *in vitro* by mechanisms that are currently unclear. Although similar mutations of *TNFRSF11A* have been excluded as a cause of classical PDB,[26] there is

strong evidence to suggest that common variants at the *TNFRSF11A* locus predispose to the disease.[27–29]

Mutations in the *TNFRSF11B* gene encoding osteoprotegerin (OPG) are the cause of juvenile Paget's disease (JPD). This is a rare disorder associated with grossly abnormal bone remodeling, bone expansion, and bone deformity which presents in childhood and adolescence. Various mutations have been described in JPD including deletions involving the whole gene[30] and various missense mutations.[31] These are loss-of-function mutations that result in the OPG protein not being produced at all or that result in formation of an abnormal OPG protein, which is incapable of binding to RANKL and inhibiting bone resorption.[32] Mutations of *TNFRSF11B* have not so far been detected in classical PDB, but there is some evidence to suggest that variants at the *TNFRSF11B* locus predispose to PDB at least in women.[33,34]

Mutations in the *VCP* gene have been identified as the cause of IBMPFD, a syndrome characterized by myopathy, and dementia, which is often accompanied by PDB.[35] The predominant clinical feature of this syndrome is myopathy, which typically presents after the age of 40 and is observed in 90% of patients. About 43% of patients also develop typical PDB lesions and 37% develop dementia.[23] All of the mutations affect highly conserved amino acid residues clustered in the N-terminal domain of the *VCP* gene product, which is known to be involved in ubiquitin binding. This finding is of interest in relation to the fact the *VCP* is known to regulate degradation of the IKB-α protein, which is involved in NF-κB signaling.[36] Mutations of *VCP* have been excluded as a cause of classical PDB.[37]

Genes that cause classical Paget's disease

SQSTM1

Mutations affecting the *SQSTM1* gene cause a high penetrance form of PDB, which is inherited in an autosomal dominant manner. The causal mutations were identified as the result of a positional cloning approach following the identification of a strong susceptibility locus for PDB on chromosome 5q35 in two independent populations.[14,15] Mutation screening of genes within the region identified a proline to leucine mutation at codon 392 of the *SQSTM1* gene as the cause of 5q35 linked PDB in the French Canadian population.[38] Soon after this, additional mutations of *SQSTM1* clustering in the ubiquitin associated (UBA) domain were identified in British patients.[39] A large number of *SQSTM1* mutations have now been identified in PDB and most of these affect the UBA domain.[40,41] The *SQSTM1* gene encodes p62, which is an adaptor protein in the NF-κB signaling pathway.[42] In addition to its role in regulating signaling downstream of the RANK, TNF, and NGF receptors,[43] p62 appears to play a key role in regulating other cellular processes through its involvement in autophagy.[44–46] Mutations of *SQSTM1* occur in about 40% of patients with a family history of PDB and up to 10% of "sporadic" cases.[18,38–40,47] The mechanisms by which *SQSTM1* mutations lead to PDB are incompletely understood, but a common feature of virtually all mutations described so far is that they interfere with the ability of p62 to bind to ubiquitin.[48] This leads to enhanced NF-κB signaling and increased sensitivity of osteoclast precursors to RANKL by mechanisms that remain incompletely understood.[48] It has previously been suggested that *SQSTM1* plays a permissive, rather than causal, role in PDB on the basis that the disease is incompletely penetrant and that mice carrying the P394L mutation of *sqstm1* (equivalent to the P392L human mutation) did not develop PDB-like bone lesions in the spine.[49] However, recent studies have shown that mice with the P394L mutation do develop a bone disorder with remarkable similarity to PDB, characterized by focal osteolytic and osteosclerotic lesions predominantly affecting the lower limbs, woven bone, and inclusion bodies very similar to those observed in the human disease.[50] The difference between these studies is most likely due to technical differences in analyzing the skeletal phenotype. In one study,[49] screening for lesions was limited to analysis of the lumbar spine and was carried out by histology, whereas in another study,[50] screening for lesions was done by MicroCT and included analysis of the spine and lower limbs.

Genome-wide significant loci for classical PDB

Recently, additional variants that predispose to PDB have been identified by genome-wide association studies (GWAS), which showed that single nucleotide polymorphisms at the *CSF1*, *OPTN*, *TNFRSF11A*, *TM7SF4*, *NUP205*, *PML*, and *RIN3* loci were significant risk factors for the development of PDB [27,29] (Table 1). The *CSF1*, *TNFRSF11A*, and *TM7SF4* loci contain strong functional candidate genes for PDB susceptibility. The *CSF1* gene is situated on chromosome 1p13 and encodes macrophage colony stimulating factor (M-CSF), a cytokine that is essential for osteoclast and macrophage differentiation.[51] The *TM7SF4* gene is situated on chromosome 8q22 and encodes dendritic cell–specific transmembrane protein (DC-STAMP), a cell surface protein that is essential for the formation of multinucleated osteoclasts and macrophage polykaryons.[52] The *TNFRSF11A* gene is situated on chromosome 18q21 and encodes RANK, a receptor that is essential for osteoclast differentiation and bone resorption.[53] The remaining four loci contain genes that have not previously been implicated in bone metabolism. The 7q33 locus contains three genes (*CNOT4*, *NUP205*, and *SLC13A4*) and two predicted coding transcripts (*PL-5283* and *FAM180A*). Any could be responsible for the association observed, but the strongest signal was within *NUP205*, which encodes the nucleoporin 205kd protein, a component of the nuclear pore complex.[54] The only

Table 1. Genome-wide loci for susceptibility to PDB

Locus	Nearest gene(s)	SNP	*P* value	Odds ratio (95% CI)
1p13	*CSF1*	rs10494112	7.06×10^{-35}	1.72 (1.57–1.87)
7q33	*CNOT4, NUP205, SLC13A4*	rs4294134	8.45×10^{-10}	1.45 (1.29–1.63)
8q22	*TM7SF4*	rs2458413	7.38×10^{-17}	1.40 (1.29–1.51)
10p13	*OPTN*	rs1561570	4.37×10^{-38}	1.67 (1.54–1.80)
14q32	*RIN3*	rs10498635	2.55×10^{-11}	1.44 (1.29–1.60)
15q24	*PML, GOLGA6A*	rs5742915	1.6×10^{-14}	1.34 (1.25–1.45)
18q21	*TNFRSF11A*	rs3018362	7.98×10^{-21}	1.45 (1.34–1.56)

The *P* values shown and odds ratios for association with PDB are from the strongest associated SNP in each locus. Data are from Albagha *et al.*[27]

gene that lies within the 10p13 locus is *OPTN*, which encodes optineurin; a protein involved in regulating NF-κB signaling and autophagy.[55,56] Mutations in *OPTN* have been described in glaucoma[57] and motor neurone disease,[58] but the role of *OPTN* in bone is unknown. It seems likely that the causal genetic variants at the *OPTN* locus may have a large effect on susceptibility to PDB because the 10p13 region was also identified as a candidate locus for PDB by linkage studies in families with autosomal dominant inheritance of PDB.[59] However, defining the molecular mechanisms by which variants at this locus predispose to PDB will require further studies.

The most likely candidate gene for PDB susceptibility within the 14q32 locus is *RIN3*, which encodes Ras and Rab interactor 3 protein. The RIN3 protein plays a role in vesicular trafficking by interacting with small GTPases[60] and could conceivably affect osteoclast function through this mechanism, but its role in bone cell function has not been studied. Two candidate genes are contained within the 15q24 locus; *PML* and *GOLGA6A*. The strongest associated SNP lies within *PML* and codes for a phenylalanine to leucine change at amino acid 645. The PML protein is involved in regulating cell growth, apoptosis, and senescence[61] and also has been shown to regulate TGF-β signaling.[62] However, PML has not previously been implicated in the regulation of bone metabolism. The other gene within this locus encodes a member of the golgin family of proteins, which are thought to play a role in membrane fusion and as structural supports for the Golgi cisternae. Mutations in other members of the golgin family have been shown to cause a lethal skeletal dysplasia[63] and the syndrome of Gerodermia osteodysplastica, which is characterized by abnormal skin and severe osteoporosis.[64] The role of *GOLGA6A* gene in bone has not yet been studied.

The seven genes identified from the GWAS studies mentioned earlier showed independent association with PDB consistent with a multiplicative model for susceptibility. Indeed, together the identified genetic variants accounted for ∼86% of the population-attributable risk of PDB.[29] In this context, the population-attributable risk provides an estimate of the proportion of PDB cases that are associated with carriage of the risk alleles described earlier, taking into account their frequency in the population and combined effect size. Furthermore,

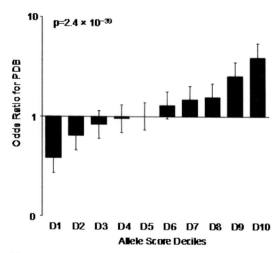

Figure 1. Common genetic variants increase the risk of PDB substantially. The risk of PDB is increased with an increasing number of risk alleles carried by individuals. Individuals who carried the greatest number of risk alleles (D1) had a 10-fold increase in risk when compared with those who carried the lowest number of risk alleles (D10). Redrawn from Albagha *et al.*[27]

the risk of developing PDB increased with the number of risk alleles carried, such that patients carrying the greatest number of risk alleles had a 10-fold increase in PDB risk compared to those carrying the smallest number (Fig. 1).

Other candidate genes for PDB

In addition to the genes and loci mentioned earlier, polymorphisms in various other candidate genes and loci have been studied in patients with PDB. These have resulted in the identification of nominally significant associations between PDB and polymorphisms in the *CASR*,[65] *ESR1*,[65] and *TNFRSF11B*[34,66] genes. A positive association has been observed in one study between polymorphisms in the *VCP* gene and PDB,[67] whereas in another study, no association was observed.[37] Other candidate genes that have been studied with negative results include, *IL1B*, *IL1RA*, *IL6*, *IL8*, *TNFA*, *TNFSF11*, and *VDR*.[65,68,69] The association between *TNFRSF11B* polymorphisms and PDB is of some interest because current evidence suggests that these variants predispose to PDB only in women. Because *TNFRSF11B* did not emerge as a candidate locus for PDB overall in recent GWAS studies,[27,29] further analysis of this locus as a determinant of PDB in women would be of interest.

Environmental factors and PDB

The reducing prevalence of PDB in some countries over recent years and delayed onset in offspring of patients who carry *SQSTM1* mutations[10] indicate that environmental triggers also play a significant role in regulating disease occurrence and severity.[9] Viral infection was the first suggested environmental trigger of PDB rooted in the observation of inclusion bodies in osteoclasts from affected patients that were thought to be viral nucleocapsids.[70] This led to the suggestion that PDB may be caused by slow paramyxovirus infection of osteoclast precursors. Subsequent studies into the role of viral infection in PDB have been inconclusive however with some studies reporting positive results[71–73] and others reporting negative results.[74–78,79] The most comprehensive study investigating this issue showed no evidence of paramyxovirus transcripts in PDB patients and indicated that PCR contamination could be the reason for previously reported positive findings.[80] Experimental studies have shown that paramyxoviruses enhance osteoclast formation *in vitro*[81] and that bone turnover is enhanced in mice overexpressing measles virus nucleocapsid protein in osteoclasts.[82] However, these effects are not specific to paramyxoviruses because other viral proteins, such as HLTV1 Tax, have been found to enhance bone turnover in mice. Other possible environmental triggers, such as childhood dietary calcium intake,[83] mechanical loading of the skeleton,[84] zoonotic infections,[85] and pesticides and toxins[86] have also been suggested as triggers for PDB. The evidence supporting these observations is drawn from isolated reports or anecdotal observations, however, and further studies will be required to confirm or refute these observations.

Conclusion

There have been tremendous advances in knowledge and understanding of the pathogenesis of PDB over the past decade, driven to a large extent by the identification of genes that predispose to the disease. Although these studies have shown that genetic factors play a key role in the pathogenesis of PDB, the molecular and cellular mechanisms by which these genes cause the disease remain incompletely understood. Further work will therefore be required to identify the causal genetic variants; the cellular and molecular mechanisms by which they regulate bone cell function; and to determine how these genetic factors interact with environmental triggers to influence the occurrence and severity of this fascinating disease.

Conflicts of interest

The authors declare no conflicts of interest.

References

1. Kanis, J.A. 1992. *Pathophysiology and Treatment of Paget's Disease of Bone*. Vol. 1: 1–293. Martin Dunitz. London.
2. van Staa, T.P., P. Selby, H.G. Leufkens, *et al.* 2002. Incidence and natural history of Paget's disease of bone in England and Wales. *J. Bone Miner. Res.* **17:** 465–471.
3. Detheridge, F.M., P.B. Guyer & D.J. Barker. 1982. European distribution of Paget's disease of bone. *Br. Med. J.* **285:** 1005–1008.
4. Barker, D.J. 1984. The epidemiology of Paget's disease of bone. [Review]. *Br. Med. Bull.* **40:** 396–400.
5. Takata, S., J. Hashimoto, K. Nakatsuka, *et al.* 2006. Guidelines for diagnosis and management of Paget's disease of bone in Japan. *J. Bone Miner. Metab* **24:** 359–367.
6. Joshi, S.R., S. Ambhore, N. Butala, *et al.* 2006. Paget's disease from Western India. *J. Assoc. Physicians India* **54:** 535–538.
7. Poor, G., J. Donath, B. Fornet & C. Cooper. 2006. Epidemiology of Paget's disease in Europe: the prevalence is decreasing. *J. Bone Miner. Res.* **21:** 1545–1549.
8. Cooper, C., K. Schafheutle, E. Dennison, *et al.* 1999. The epidemiology of Paget's disease in Britain: is the prevalence decreasing? *J. Bone Miner. Res.* **14:** 192–197.
9. Cundy, H.R., G. Gamble, D. Wattie, *et al.* 2004. Paget's disease of bone in New Zealand: continued decline in disease severity. *Calcif. Tissue Int.* **75:** 358–364.
10. Bolland, M.J., P.C. Tong, D. Naot, *et al.* 2007. Delayed development of Paget's disease in offspring inheriting SQSTM1 mutations. *J. Bone Miner. Res.* **22:** 411–415.
11. Gennari, L., S.M. Di, D. Merlotti, *et al.* 2005. Prevalence of Paget's disease of bone in Italy. *J. Bone Miner. Res.* **20:** 1845–1850.
12. Tiegs, R.D., C.M. Lohse, P.C. Wollan & L.J. Melton. 2000. Long-term trends in the incidence of Paget's disease of bone. *Bone* **27:** 423–427.
13. Morales-Piga, A.A., J.S. Rey-Rey, J. Corres-Gonzalez, *et al.* 1995. Frequency and characteristics of familial aggregation of Paget's disease of bone. *J. Bone Miner. Res.* **10:** 663–670.
14. Hocking, L.J., C.A. Herbert, R.K. Nicholls, *et al.* 2001. Genomewide search in familial Paget disease of bone shows evidence of genetic heterogeneity with candidate loci on chromosomes 2q36, 10p13, and 5q35. *Am. J. Hum. Genet.* **69:** 1055–1061.
15. Laurin, N., J.P. Brown, A. Lemainque, *et al.* 2001. Paget disease of bone: mapping of two loci at 5q35-qter and 5q31. *Am. J. Hum. Genet.* **69:** 528–543.
16. Siris, E.S., R. Ottman, E. Flaster & J.L. Kelsey. 1991. Familial aggregation of Paget's disease of bone. *J. Bone Miner. Res.* **6:** 495–500.

17. Sofaer, J.A., S.M. Holloway & A.E. Emery. 1983. A family study of Paget's disease of bone. *J. Epidemiol. Community Health* **37:** 226–231.

18. Eekhoff, E.W., M. Karperien, D. Houtsma, *et al.* 2004. Familial Paget's disease in The Netherlands: occurrence, identification of new mutations in the sequestosome 1 gene, and their clinical associations. *Arthr. Rheum.* **50:** 1650–1654.

19. Osterberg, P.H., R.G. Wallace, D.A. Adams, *et al.* 1988. Familial expansile osteolysis: a new dysplasia. *J. Bone Joint Surg. Br.* **70:** 255–260.

20. Whyte, M.P., B.G. Mills, W.R. Reinus, *et al.* 2000. Expansile skeletal hyperphosphatasia: a new familial metabolic bone disease. *J. Bone Miner. Res.* **15:** 2330–2344.

21. Nakatsuka, K., Y. Nishizawa & S.H. Ralston. 2003. Phenotypic characterization of early onset Paget's disease of bone caused by a 27-bp duplication in the TNFRSF11A gene. *J. Bone Miner. Res.* **18:** 1381–1385.

22. Whyte, M.P. & A.E. Hughes. 2002. Expansile skeletal hyperphosphatasia is caused by a 15-base pair tandem duplication in TNFRSF11A encoding RANK and is allelic to familial expansile osteolysis. *J. Bone Miner. Res.* **17:** 26–29.

23. Kovach, M.J., B. Waggoner, S.M. Leal, *et al.* 2001. Clinical delineation and localization to chromosome 9p13.3-p12 of a unique dominant disorder in four families: hereditary inclusion body myopathy, Paget disease of bone, and frontotemporal dementia. *Mol. Genet. Metab* **74:** 458–475.

24. Hughes, A.E., S.H. Ralston, J. Marken, *et al.* 2000. Mutations in TNFRSF11A, affecting the signal peptide of RANK, cause familial expansile osteolysis. *Nat. Genet.* **24:** 45–48.

25. Crockett, J.C., D.J. Mellis, K.I. Shennan, *et al.* 2011. Signal peptide mutations in rank prevent downstream activation of NFkappaB. *J. Bone Miner. Res.* **26:** 1926–1938.

26. Wuyts, W., L. Van Wesenbeeck, A. Morales-Piga, *et al.* 2001. Evaluation of the role of RANK and OPG genes in Paget's disease of bone. *Bone* **28:** 104–107.

27. Albagha, O.M.E., S. Wani, M.R. Visconti, *et al.* 2011. Genome-wide association identifies three new susceptibility loci for Paget's disease of bone. *Nat. Genet.* **43:** 685–689.

28. Chung, P.Y., G. Beyens, P.L. Riches, *et al.* 2010. Genetic variation in the TNFRSF11A gene encoding RANK is associated with susceptibility to Paget's disease of bone. *J. Bone Miner. Res.* **25:** 2316–2329.

29. Albagha, O.M., M.R. Visconti, N. Alonso, *et al.* 2010. Genome-wide association study identifies variants at CSF1, OPTN and TNFRSF11A as genetic risk factors for Paget's disease of bone. *Nat. Genet.* **42:** 520–524.

30. Whyte, M.P., S.E. Obrecht, P.M. Finnegan, *et al.* 2002. Osteoprotegerin deficiency and juvenile Paget's disease. *N. Engl. J. Med.* **347:** 175–184.

31. Cundy, T., M. Hegde, D. Naot, *et al.* 2002. A mutation in the gene TNFRSF11B encoding osteoprotegerin causes an idiopathic hyperphosphatasia phenotype. *Hum. Mol. Genet.* **11:** 2119–2127.

32. Middleton-Hardie, C., Q. Zhu, H. Cundy, *et al.* 2006. Deletion of aspartate 182 in OPG causes juvenile Paget's disease by impairing both protein secretion and binding to RANKL. *J. Bone Miner. Res.* **21:** 438–445.

33. Beyens, G., A. Daroszewska, F. De Freitas, *et al.* 2007. Identification of sex specific effects of TNFRSF11B polymorphisms on the risk of Paget's disease of bone. *Calcif. Tissue Int.* **80:** S42.

34. Daroszewska, A., L.J. Hocking, F.E.A. McGuigan, *et al.* 2004. Susceptibility to Paget's disease of bone is influenced by a common polymorphic variant of Osteoprotegerin. *J. Bone Miner. Res.* **19:** 1506–1511.

35. Watts, G.D., J. Wymer, M.J. Kovach, *et al.* 2004. Inclusion body myopathy associated with Paget disease of bone and frontotemporal dementia is caused by mutant valosin-containing protein. *Nat. Genet.* **36:** 377–381.

36. Dai, R.M., E. Chen, D.L. Longo, *et al.* 1998. Involvement of valosin-containing protein, an ATPase Co-purified with IkappaBalpha and 26 S proteasome, in ubiquitin-proteasome-mediated degradation of IkappaBalpha. *J. Biol. Chem.* **273:** 3562–3573.

37. Lucas, G.J., S.G. Mehta, L.J. Hocking, *et al.* 2006. Evaluation of the role of Valosin-containing protein in the pathogenesis of familial and sporadic Paget's disease of bone. *Bone* **38:** 280–285.

38. Laurin, N., J.P. Brown, J. Morissette & V. Raymond. 2002. Recurrent mutation of the gene encoding sequestosome 1 (SQSTM1/p62) in Paget disease of bone. *Am. J. Hum. Genet.* **70:** 1582–1588.

39. Hocking, L.J., G.J.A. Lucas, A. Daroszewska, *et al.* 2002. Domain specific mutations in Sequestosome 1 (SQSTM1) cause familial and sporadic Paget's disease. *Hum. Mol. Genet.* **11:** 2735–2739.

40. Beyens, G., H.E. Van, D.K. Van, *et al.* 2004. Evaluation of the role of the SQSTM1 gene in sporadic Belgian patients with Paget's disease. *Calcif. Tissue Int.* **75:** 144–152.

41. Falchetti, A., S.M. Di, F. Marini, *et al.* 2009. Genetic epidemiology of Paget's disease of bone in Italy: sequestosome1/p62 gene mutational test and haplotype analysis at 5q35 in a large representative series of sporadic and familial Italian cases of Paget's disease of bone. *Calcif. Tissue Int.* **84:** 20–37.

42. Layfield, R. & L.J. Hocking. 2004. SQSTM1 and Paget's disease of bone. *Calcif. Tissue Int.* **75:** 347–357.

43. Duran, A., M. Serrano, M. Leitges, *et al.* 2004. The atypical PKC-interacting protein p62 is an important mediator of RANK-activated osteoclastogenesis. *Dev. Cell* **6:** 303–309.

44. Kirkin, V., T. Lamark, J. Johansen & I. Dikic. 2009. NBR1 cooperates with p62 in selective autophagy of ubiquitinated targets. *Autophagy* **5:** 732–733.

45. Kirkin, V., T. Lamark, Y.S. Sou, *et al.* 2009. A role for NBR1 in autophagosomal degradation of ubiquitinated substrates. *Mol. Cell* **33:** 505–516.

46. Komatsu, M., S. Waguri, M. Koike, *et al.* 2007. Homeostatic levels of p62 control cytoplasmic inclusion body formation in autophagy-deficient mice. *Cell* **131:** 1149–1163.

47. Visconti, M.R., A.L. Langston, N. Alonso, *et al.* 2010. Mutations of SQSTM1 are associated with severity and clinical outcome in Paget's disease of bone. *J. Bone Miner. Res.* **25:** 2368–2373.

48. Cavey, J.R., S.H. Ralston, P.W. Sheppard, *et al.* 2006. Loss of ubiquitin binding is a unifying mechanism by which mutations of SQSTM1 cause Paget's disease of bone. *Calcif. Tissue Int.* **78:** 271–277.

49. Hiruma, Y., N. Kurihara, M.A. Subler, *et al.* 2008. A SQSTM1/p62 mutation linked to Paget's disease increases the osteoclastogenic potential of the bone microenvironment. *Hum. Mol. Genet.* **17:** 3708–3719.

50. Daroszewska, A., R.J. van't Hof, J.A. Rojas, *et al.* 2011. A point mutation in the ubiquitin associated domain of SQSMT1 is sufficient to cause a Paget's disease like disorder in mice. *Hum. Mol. Genet.* **20:** 2734–2744.

51. Yoshida, H., S. Hayashi, T. Kunisada, *et al.* 1990. The murine mutation osteopetrosis is in the coding region of the macrophage colony stimulating factor gene. *Nature* **345:** 442–444.

52. Yagi, M., T. Miyamoto, Y. Sawatani, *et al.* 2005. DC-STAMP is essential for cell-cell fusion in osteoclasts and foreign body giant cells. *J. Exp. Med.* **202:** 345–351.

53. Li, J., I. Sarosi, X.Q. Yan, *et al.* 2000. RANK is the intrinsic hematopoietic cell surface receptor that controls osteoclastogenesis and regulation of bone mass and calcium metabolism. *Proc. Natl Acad. Sci. USA* **97:** 1566–1571.

54. Grandi, P., T. Dang, N. Pane, *et al.* 1997. Nup93, a vertebrate homologue of yeast Nic96p, forms a complex with a novel 205-kDa protein and is required for correct nuclear pore assembly. *Mol. Biol. Cell* **8:** 2017–2038.

55. Zhu, G., C.J. Wu, Y. Zhao & J.D. Ashwell. 2007. Optineurin negatively regulates TNFalpha-induced NF-kappaB activation by competing with NEMO for ubiquitinated RIP. *Curr. Biol.* **17:** 1438–1443.

56. Shen, X., H. Ying, Y. Qiu, *et al.* 2011. Processing of optineurin in neuronal cells. *J. Biol. Chem.* **286:** 3618–3629.

57. Rezaie, T., A. Child, R. Hitchings, *et al.* 2002. Adult-onset primary open-angle glaucoma caused by mutations in optineurin. *Science* **295:** 1077–1079.

58. Maruyama, H., H. Morino, H. Ito, *et al.* 2010. Mutations of optineurin in amyotrophic lateral sclerosis. *Nature* **465:** 223–226.

59. Lucas, G., P. Riches, L. Hocking, *et al.* 2008. Identification of a major locus for Paget disease on chromosome 10p13 in families of British descent. *J. Bone Miner. Res.* **23:** 58–63.

60. Kajiho, H., K. Saito, K. Tsujita, *et al.* 2003. RIN3: a novel Rab5 GEF interacting with amphiphysin II involved in the early endocytic pathway. *J. Cell Sci.* **116:** 4159–4168.

61. Salomoni, P. & P.P. Pandolfi. 2002. The role of PML in tumor suppression. *Cell* **108:** 165–170.

62. Lin, H.K., S. Bergmann & P.P. Pandolfi. 2004. Cytoplasmic PML function in TGF-beta signalling. *Nature* **431:** 205–211.

63. Smits, P., A.D. Bolton, V. Funari, *et al.* 2010. Lethal skeletal dysplasia in mice and humans lacking the golgin GMAP-210. *N. Engl. J. Med.* **362:** 206–216.

64. Hennies, H.C., U. Kornak, H. Zhang, *et al.* 2008. Gerodermia osteodysplastica is caused by mutations in SCYL1BP1, a Rab-6 interacting golgin. *Nat. Genet.* **40:** 1410–1412.

65. Donath, J., G. Speer, G. Poor, *et al.* 2004. Vitamin D receptor, oestrogen receptor-alpha and calcium-sensing receptor genotypes, bone mineral density and biochemical markers in Paget's disease of bone. *Rheumatology (Oxford)* **43:** 692–695.

66. Beyens, G., A. Daroszewska, F.F. de Freitas, *et al.* 2007. Identification of sex-specific associations between polymorphisms of the osteoprotegerin gene, TNFRSF11B, and Paget's disease of bone. *J. Bone Miner. Res.* **22:** 1062–1071.

67. Chung, P.Y., G. Beyens, F.F. de Freitas, *et al.* 2011. Indications for a genetic association of a VCP polymorphism with the pathogenesis of sporadic Paget's disease of bone, but not for TNFSF11 (RANKL) and IL-6 polymorphisms. *Mol. Genet. Metab* **103:** 287–292.

68. Corral-Gudino, L., J. Del Pino-Montes, J. Garcia-Aparicio, *et al.* 2010. Paget's disease of bone is not associated with common polymorphisms in interleukin-6, interleukin-8 and tumor necrosis factor alpha genes. *Cytokine* **52:** 146–150.

69. Corral-Gudino, L., J. Del Pino-Montes, J. Garcia-Aparicio, *et al.* 2006. -511 C/T IL1B gene polymorphism is associated to resistance to bisphosphonates treatment in Paget disease of bone. *Bone* **38:** 589–594.

70. Rebel, A., K. Malkani, M. Basle, *et al.* 1974. Ultrastructural characteristics of osteoclasts in Paget's disease. *Rev. Rhum. Mal. Osteoartic.* **41:** 767–771.

71. Gordon, M.T., A.P. Mee, D.C. Anderson & P.T. Sharpe. 1992. Canine distemper transcripts sequenced from pagetic bone. *Bone Miner.* **19:** 159–174.

72. Gordon, M.T., D.C. Anderson & P.T. Sharpe. 1991. Canine distemper virus localised in bone cells of patients with Paget's disease. *Bone* **12:** 195–201.

73. Mee, A.P., J.A. Dixon, J.A. Hoyland, *et al.* 1998. Detection of Canine Distemper Virus in 100% of Paget's disease samples by in situ-reverse transcriptase polymerase chain reaction. *Bone* **23:** 171–175.

74. Birch, M.A., W. Taylor, W.D. Fraser, *et al.* 1994. Absence of paramyxovirus RNA in cultures of pagetic bone cells and in pagetic bone. *J. Bone Miner. Res.* **9:** 11–16.

75. Helfrich, M.H., R.P. Hobson, P.S. Grabowski, *et al.* 2000. A negative search for a paramyxoviral etiology of Paget's disease of bone: molecular, immunological, and ultrastructural studies in UK patients. *J. Bone Miner. Res.* **15:** 2315–2329.

76. Nuovo, M.A., G.J. Nuovo, P. MacConnell, *et al.* 1992. In situ analysis of Paget's disease of bone for measles-specific PCR-amplified cDNA. *Diagn. Mol. Pathol.* **1:** 256–265.

77. Ooi, C.G., C.A. Walsh, J.A. Gallagher & W.D. Fraser. 2000. Absence of measles virus and canine distemper virus transcripts in long-term bone marrow cultures from patients with Paget's disease of bone. *Bone* **27:** 417–421.

78. Ralston, S.H., F.S. DiGiovine, S.J. Gallacher, *et al.* 1991. Failure to detect paramyxovirus sequences in Paget's disease of bone using the polymerase chain reaction. *J. Bone Miner. Res.* **6:** 1243–1248.

79. Matthews, B.G., M.A. Afzal, P.D. Minor, *et al.* 2008. Failure to detect measles virus RNA in bone cells from patients with Paget's disease. *J. Clin. Endocrinol. Metab.* **93:** 1398–1401.

80. Ralston, S.H., M.A. Afzal, M.H. Helfrich, *et al.* 2007. Multicenter blinded analysis of RT-PCR detection methods for paramyxoviruses in relation to Paget's disease of bone. *J. Bone Miner. Res.* **22:** 569–577.

81. Selby, P.L., M. Davies & A.P. Mee. 2006. Canine distemper virus induces human osteoclastogenesis through NF-kappaB

and sequestosome 1/P62 activation. *J. Bone Miner. Res.* **21:** 1750–1756.

82. Kurihara, N., H. Zhou, S.V. Reddy, *et al.* 2006. Expression of measles virus nucleocapsid protein in osteoclasts induces Paget's disease-like bone lesions in mice. *J. Bone Miner. Res.* **21:** 446–455.

83. Siris, E.S. 1994. Epidemiological aspects of Paget's disease: family history and relationship to other medical conditions. *Semin. Arthritis Rheum.* **23:** 222–225.

84. Solomon, L.R. 1979. Billiard-player's fingers: an unusual case of Paget's disease of bone. *Br. Med. J.* **1:** 931.

85. Lopez-Abente, G., A. Morales-Piga, A. Elena-Ibanez, *et al.* 1997. Cattle, pets, and Paget's disease of bone. *Epidemiology* **8:** 247–251.

86. Lever, J.H. 2002. Paget's disease of bone in Lancashire and arsenic pesticide in cotton mill wastewater: a speculative hypothesis. *Bone* **31:** 434–436.

Ann. N.Y. Acad. Sci. ISSN 0077-8923

Recent progress in understanding molecular mechanisms of cartilage degeneration during osteoarthritis

Meina Wang,[1] Jie Shen,[1] Hongting Jin,[1,2] Hee-Jeong Im,[3] John Sandy,[3] and Di Chen[1,3]

[1]Center for Musculoskeletal Research, University of Rochester Medical Center, Rochester, New York. [2]Institute of Orthopaedics and Traumatology, Zhejiang Chinese Medical University, Hangzhou, Zhejiang Province, China. [3]Department of Biochemistry, Rush University Medical Center, Chicago, Illinois

Address for correspondence: Di Chen, MD, PhD, Department of Biochemistry, Rush University Medical Center, 1735 West Harrison Street, Cohn Research Building, Ste. 508, Chicago, IL 60612-3823. di_chen@rush.edu

Osteoarthritis (OA) is a highly prevalent disease affecting more than 20% of American adults. Predispositions include joint injury, heredity, obesity, and aging. Biomechanical alterations are commonly involved. However, the molecular mechanisms of this disease are complex, and there is currently no effective disease-modifying treatment. The initiation and progression of OA subtypes is a complex process that at the molecular level probably involves many cell types, signaling pathways, and changes in extracellular matrix. *Ex vivo* studies with tissue derived from OA patients and *in vivo* studies with mutant mice have suggested that pathways involving receptor ligands such as TGF-β1, WNT3a, and Indian hedgehog; signaling molecules such as Smads, β-catenin, and HIF-2a; and peptidases such as MMP13 and ADAMTS4/5 are probably involved to some degree. This review focuses on molecular mechanisms of OA development related to recent findings.

Keywords: osteoarthritis; articular cartilage; growth factors; Mmp13; Adamts5

Normal cartilage degeneration in osteoarthritis

During the past several decades, there has been an explosion of reports describing abnormalities in the gross appearance, material properties, cellular morphologies, biochemical composition, and gene expression in articular cartilages from humans to animals with osteoarthritis (OA)-like joint pathology.[1-7] Notwithstanding this vast research enterprise, the pathogenetic mechanisms involved in the initiation and progression of OA resulting from one or more of the many predisposing factors (e.g., age, injury, genetics, and obesity) remain unclear. A major consideration has been whether the different predispositions translate into a "final common pathway" in the articular cartilage[8] that might be amenable to therapeutic intervention.

OA is the most common joint disorder worldwide and is a major cause of disability. There are currently no treatments capable of markedly altering its progression. Characteristic features of OA include phenotypic changes in the cells of the superficial layer of the articular cartilage (AC), chondrocyte hypertrophy and apoptosis, progressive fibrillation of the AC, subchondral bone sclerosis, osteophyte formation, and increased remodeling of the periarticular bone.[4,6] The articular cartilage has received much of the attention in OA studies because gross articular cartilage damage is the most obvious pathologic feature leading to joint dysfunction. AC is a smooth, lubricated, reversibly compressible tissue that protects the underlying bones from biomechanical damage during joint loading. About 75% of the wet weight of AC is water, and about 70% of the dry weight is collagen. The principal collagens of adult articular cartilage are type II (often present as a heteropolymer with types IX and XI), type III, and a small amount of types V, VI, and X.[9,10] The noncollagenous matrix (about 20% of the dry weight) is mostly the proteoglycan aggrecan, which is present largely in link-stabilized aggregates with hyaluronan (HA). Full-length aggrecan is itself about 10% (w/w) core protein and 90% (w/w)

doi: 10.1111/j.1749-6632.2011.06258.x

chondroitin sulfate (CS). Other cartilage proteoglycans, many involved in controlling collagen fibril formation and pericellular matrix organization, include decorin, biglycan, fibromodulin, lumican, epiphycan, and perlecan. However, the relative abundance of these has not been accurately determined. It is important to note that the abundance and composition of all cartilage components can vary with tissue depth,[11] maturation and aging,[12] and diseases such as OA.[13]

The chondrocytic cells of the articular cartilage are organized into three layers—superficial, middle, and deep—where they represent about 2.5%, 2%, and 1.5% of the cartilage volume, respectively.[14] The cartilage collagens form a dense fibrous meshwork that constrains the highly concentrated aggrecan, which in turn retains water due to the osmotic effect of its negatively charged chondroitin sulfate chains. The chondrocytic cells, which are embedded in these matrix networks, produce and maintain the cartilage by synthesizing and degrading matrix components in response to environmental cues such as growth factors, cytokines, and biomechanical change. Mature articular cartilage is a product of postnatal remodeling of the cartilaginous epiphysis. Development begins with the aggregation of mesenchymal precursors and differentiation of the cells into chondrocytes, as indicated by expression of Sox-5, -6, and -9,[15] and the secretion of matrix components, such as collagens II, VI, IX, and XI; link protein; and the hyaluronan (HA)-binding proteoglycans, aggrecan, and versican. Chondrocytes present in this cartilaginous anlage of the developing skeleton, subsequently organize into zones of quiescence and proliferation. Groups of proliferative cells form proliferating zone columns wherein the cells undergo a differentiation program through prehypertrophy and hypertrophy. Hypertrophic cells are characterized by a high expression of markers such as Runx2, collagen X, and alkaline phosphatase. These changes in expression are accompanied by matrix calcification and the emergence of cells expressing markers such as VEGF and osteocalcin, which in turn results in vascular invasion, chondrocyte apoptosis, and trabecular bone deposition.

In contrast to the proliferative cells, the quiescent chondrocytes of the original cartilage template are the source of the mature articular cartilage. This tissue is characterized by flattened "fibroblastic" cells in the surface zone and small groups of more rounded "chondroid" cells in the mid- and deep zones. The composition and organization of the matrix in each zone is different, indicating that the maintenance of articular cartilage relies on zone-specific programs for the synthesis and turnover of each matrix component. It also appears that the superficial zone[16] and the deep zone[17] are a source of progenitor cells, which are needed to replace chondrocytes lost by apoptosis, necrosis, or autophagy under biomechanical or biologic stress.

Chondrocyte activities related to the induction and/or progression of OA

Many of the foundational studies on mechanisms of cartilage degradation were performed with normal cartilage or chondrocytes. The source of tissue has commonly been animals (lapine, bovine, porcine) or normal human cartilage taken postmortem or at amputation. These analyses identified a range of ligands (e.g., cytokines and growth factors) and receptors (e.g, IL-1R and TGF-βRII) that alter chondrocyte-mediated matrix turnover and the gene expression of effector molecules (e.g., peptidases). Reports have focused on cell responses that are consistent with, but clearly cannot establish alone, a role in OA pathogenesis. Examples of this type of study are described in the series of articles[18] on the effects of mediators on peptidase expression and activity in cartilage explants. Indeed, the discovery of cartilage aggrecanase activity was made in normal cartilage explants.[19–21]

These studies provided an information base to examine changes that have occurred in the OA joint *in vivo* and that can be detected *ex vivo*. In this approach, because chondrocyte and matrix changes result from OA pathology *in vivo*, these data would appear to provide a higher level of confidence in its relevance to important aspects of disease mechanisms. Examples of this type of study are found in the detailed analysis of gene expression in cartilages obtained from different regions of the human OA joint removed at arthroplasty.[22–24] Most recently, studies done *in vivo* with genetically modified mice[25–27] have proven particularly informative in the elucidation of chondrocyte changes, which appear to have high relevance to OA pathogenesis. In the present review, we focus on *in vivo* data of OA animal models. Our objective is to optimize the likelihood that the data reviewed and summarized will have high relevance to the human disease.

TGF-β and OA

The growth factor TGF-β, which strongly inhibits articular chondrocyte hypertrophy and maturation, also represents a potential mechanism in the development of OA.[28] The intracellular signaling initiating by TGF-β is mediated through TGF-β type II and type I transmembrane Ser/Thr kinase receptors. TGF-β first binds to type II receptor, leading to the recruitment of type I receptor. This constitutively active type II receptor phosphorylates the GS domain of the type I receptor. Activated type I receptor phosphorylates R-Smads (Smad2 or Smad3) at a conserved SSXS motif at the C-terminus of Smad2/3. This phosphorylated Smad2/3 thus dissociates from receptor complex and forms a heteromeric complex with the common Smad, Smad4. This heteromeric Smad complex translocates and accumulates into the nucleus and associates with other DNA-binding proteins to regulate gene transcription.

In vivo studies show that loss of TGF-β signaling in mice causes an OA-like phenotype resembling OA in humans. The knee joints of transgenic mice that express the dominant negative type II TGF-β receptor (DNIIR) in skeletal tissue show chondrocyte hypertrophy at an early stage, followed by deceased proteoglycan, articular surface fibrillation and disorganization, as well as chondrocytes clusters in deeper zone of articular cartilage at late stage.[29] Correlating with the DNIIR transgenic mouse model, deletion of the Smad3 gene in mice also results in progressive articular cartilage degeneration resembling human OA.[30] In *Smad3* knockout mice, an abnormal increase in the hypertrophic chondrocyte number was seen at the early stage, followed by progressive loss of the smooth articular cartilage surface that covered with abnormally differentiated chondrocytes. In seven-month-old *Smad3* KO mice, articular surface is fibrillated, accompanied by vertical cleft, and osteophytes that vary in size are developed. Smurf2 is a negative regulator of TGF-β signaling in articular chondrocytes and promotes chondrocyte hypertrophy.[31,32] Smurf2 is highly expressed in human OA cartilage and is not present in normal cartilage. In chondrocyte-specific Smurf2 overexpression transgenic mice, TGF-β signaling is decreased, and expression of chondrocyte hypertrophic markers (*ColX* and *Mmp13*) is increased, leading to progressive articular cartilage degradation, including reduced cartilage area, fibrillation,

clefting, as well as subchondral sclerosis and osteophyte formation.[33] Moreover, we have recently generated chondrocyte-specific *Tgfbr2*-conditional knockout mice (*Tgfbr2^Col2CreER*) that also show OA-like features, including chondrocyte hypertrophy at an early stage, progressive cartilage degeneration, and chondrophyte and osteophyte formation (unpublished data). These observations are supported by the recent finding that high frequency of a single nucleotide polymorphism (SNP) of the *Smad3* gene was identified in patients with OA. This study, including a 527 patient cohort, demonstrated that the SNP of human *Smad3* gene is correlated with the incidence of hip and knee OA in patients.[34] The findings suggest that loss of TGF-β signaling represents one of the possible mechanisms in OA development.

Wnt, β-Catenin, and OA

β-catenin is a central molecule in the canonical Wnt signaling pathway, which controls multiple developmental processes in skeletal and joint development[35] and is critical for the progression of OA.[36] When Wnt binds to its receptor Frizzled and the co-receptors LRP5/6, the activity of downstream signaling proteins Dishevelled (Dsh) and Axins 1 and 2 are altered. This leads to the inactivation of Ser/Thr kinase GSK-3β, thus inhibiting the ubiquitination and degradation of β-catenin triggered by GSK-3β. β-catenin is then accumulated in the cytoplasm and translocated to the nucleus and binds to transcription factors LEF-1/TCF to regulate the transcription of downstream target genes. In the absence of Wnt ligands, cytoplasmic β-catenin binds the APC-Axin-GSK-3β degradation complex, and GSK-3β, in this complex, phosphorylates β-catenin. The E3 ubiquitin ligase β-TrCP then targets β-catenin for ubiquitination and proteasome degradation.[37]

β-catenin affects cell fate during early skeletal development. For example, overexpression of constitutively active β-catenin leads to the loss of chondrocyte phenotype characterized by loss of Sox9 and Col2 expression in chick chondrocytes.[35] Conditional inactivation of the β-catenin gene in mouse mesenchymal cells *in vivo* results in the loss of osteoblasts and ectopic chondrocyte formation in bone tissues through intramembranous and endochondral bone formation processes.[38]

During postnatal development, β-catenin plays an important role in chondrocyte proliferation, hypertrophy, and apoptosis. Studies suggest that dysregulation of Wnt/β-catenin signaling represents a possible mechanism of OA. Our recent findings demonstrated that β-catenin was upregulated in articular cartilage tissue derived from patients with OA.[36] In addition, human genetic studies revealed that patients with mutations of the *FrzB* gene have increased susceptibility to hip OA.[39–42] *FrzB* encodes the protein sFRP3, a secreted inhibitor of Wnt signaling. Mutations in *FrzB* cause activation of β-catenin signaling and abnormal chondrocyte hypertrophy.[43,44] Consistent with this finding, *FrzB* knockout mice are susceptible to chemical-induced OA.[45] In *Col2a1-Smurf2* transgenic mice, upregulation of β-catenin signaling in articular chondrocytes was also observed in addition to the reduction of TGF-β signaling. Primary articular chondrocytes isolated from *Col2a1-Smurf2* transgenic mice showed Smurf2-induced GSK-3β ubiquitination and subsequent upregulation of β-catenin protein levels.[33] Furthermore, overexpression of Wnt-induced signaling protein 1 (WISP-1) in the mouse knee joint also leads to cartilage destruction.[46] β-catenin-conditional activation mice also show that overexpression of β-catenin in articular chondrocytes, which causes abnormal articular chondrocyte maturation, results in cartilage degeneration in mice.[36]

It also appears to be relevant that the expression of *Mmp13* is significantly increased in articular cartilage in β-catenin gene-conditional activation mice. Consistent with this, our chondrocyte-specific *Mmp13*-conditional knockout mice have decelerated OA progression following meniscal-ligamentous injury (MLI). Further, to explore MMP13 inhibition as a therapeutic option for OA treatment, we injected CL82198, an MMP13 inhibitor (unpublished data), into WT mice after MLI surgery. We found that treatment with CL82198 decelerated MLI-induced OA progression, indicating that MMP13 is a critical player in the progression of OA, thus making it an attractive target for OA therapies.

Several reports indicate that a low level of β-catenin is associated with stable differentiated chondrocyte functions and that a high level of β-catenin is associated with loss of function due to dedifferentiation.[47] It remains to be determined whether the upregulation of β-catenin expression in articular cartilage tissue from OA patients[36] is a response to OA or part of the causative cascade.

Hypoxia-inducible factor-2α and OA

Hypoxia-inducible factor (HIF)-2α belongs to the basic helix-loop-helix/Per-ARNT-Sim (bHLH/PAS) domain transcription factor family.[48] HIF-2α is a heterodimeric protein that functions by dimerizing the α-subunit with the β-subunit members. Its activity is regulated by the level of oxygen. Under normoxic conditions, the proline residues on the α-subunits are hydroxylated, recognized by the von Hippel-Lindau (pVHL) tumor suppressor, an E3 ubiquitin ligase, and degraded by proteasome. Under hypoxic conditions, the HIF proteins do not undergo ubiquitination and proteasome degradation. The α-subunits translocate into the nucleus and dimerize with the constitutive β-subunits (also known as ARNT) to regulate HIF responsive genes.[49,50]

Studies show that the expression levels of HIF-2α are significantly increased in both human and mouse osteoarthritic cartilage.[51,52] *In vitro* promoter studies show that HIF-2α is a potent transactivator of OA marker genes, including Col10a1, MMP13, and VEGF.[51] Overexpression of HIF-2α by intra-articular injection of Ad-*EPAS1*, the gene encodes HIF-2α, leads to spontaneous OA development in knee joints of mice.[52] Moreover, *EPAS1*-heterozygous-deficient mice are resistant to surgery-induced OA development in knee joints of mice.[51,52] Consistent with these findings, a functional SNP study among a Japanese population indicates that SNP of the *EPAS1* promoter is associated with knee OA,[51] suggesting that enhanced transactivation of *EPAS1* in chondrocytes is associated with OA in humans. *In vitro* studies suggest that NF-κB is the upstream inducer of HIF-2α expression and mechanical stress upregulates NF-κB signaling. These observations suggest that the HIF-2α signaling pathway is also involved in OA development.

Indian hedgehog and OA

As an important signaling protein for chondrocyte growth and differentiation, the Indian hedgehog (Ihh) signaling pathway also plays a critical role during OA development. Ihh signaling functions through two transmembrane receptors: patched (Ptch) and smoothened (Smo). In the absence of Ihh ligand, Ptch binds to Smo to inhibit its function. During the activation of Ihh signaling, Ihh binds Ptc leading to the release of Smo. Smo will further activate Gli transcription factors to regulate Ihh responsive genes.

Activation of Ihh signaling is associated with human OA development. The expression of Ihh signaling proteins, including Gli1, Ptch, and hedgehog-interacting protein (HHIP), is highly upregulated in joint tissues of patients with OA accompanied by upregulation of *Adamts5, Col10a1, Runx2,* and *Mmp13*.[53] Furthermore, the expression of Gli1, Ptch, and HHIP is also upregulated in the injury-induced OA mouse model. *In vivo* studies show that transgenic mice with chondrocyte-specific overexpression of the *Gli2* or *Smo* gene have spontaneous OA development accompanied by upregulation of *Adamts5, Col10a1, Mmp13*, as well as aggrecan and Col2 degradation.[53] Histologic and radiographic analyses of these mice show a progressive worsening of articular cartilage with less Safranin O staining, thinner cartilage layer, or even completely degraded cartilage. Changes in subchondral bone in these mice are similar to those found in surgically induced OA knee joints. In contrast, either deletion of the *Smo* gene or treatment with an Ihh inhibitor attenuates the severity of injury-induced OA in mouse models.[53] Evidence shows that Ihh signaling is activated in the development of OA. Interestingly, it has been reported that Wnt/β-catenin interacts with Ihh signaling, and both β-catenin and Ihh signaling pathways are required for endochondral bone development.[54] However, whether β-catenin is upstream or downstream of Ihh signaling in chondrocytes remains to be determined.

MMP13 and OA

The central event of OA is progressive loss of articular cartilage. Human clinical and animal studies show that MMP13 plays a dominant role in the progression of cartilage degeneration. MMP13 is a collagenase with substrate specificity that targets collagen for degradation. Compared to other MMPs, MMP13 has a more restricted expression pattern in connective tissue.[55] MMP13 preferentially cleaves Col2, which is the most abundant protein in articular cartilage. It also targets aggrecan, types IV and IX collagen, gelatin, osteonectin, and perlecan in cartilage for degradation.[56] MMP13 has a much higher catalytic velocity rate than other MMPs for Col2 and gelatin, which make it the most potent peptidolytic enzyme among collagenases.[57,58]

Clinical investigations revealed that patients with articular cartilage destruction have high MMP13 expression,[59] suggesting that increased MMP13 may be a cause of cartilage degradation. *Mmp13*-deficient mice show no gross phenotypic abnormalities, and the only alteration is in growth plate architecture.[60,61] *Mmp13*-overexpressing transgenic mice developed spontaneous articular cartilage destruction characterized by excessive cleavage of Col2 and loss of aggrecan.[62] These results suggest that *Mmp13* deficiency does not affect articular cartilage development, but abnormally upregulated *Mmp13* can lead to postnatal cartilage destruction. Moreover, the expression of *Mmp13* is significantly increased in articular cartilage in β-catenin-conditional activation mice.[36] Consistent with above findings, we have preliminary data from chondrocyte-specific *Mmp13*-conditional knockout (*Mmp13* cKO) mice showing decelerated OA progression following meniscal-ligamentous injury (MLI) (unpublished data). To explore MMP13 inhibition as a potential therapeutic option for OA treatment, we injected CL82198, a MMP13 inhibitor, into WT mice after MLI surgery. We found that injection of CL82198 decelerated MLI-induced OA progression (unpublished data). These findings implicate that MMP13 is a critical player in the progression of OA that could serve as a molecular target for OA therapies.

ADAMTS and OA

The major contributors causing cartilage degeneration in OA are enzymes targeting collagens and aggrecan for proteolysis. In addition to MMP13, which mainly targets one of the two major structural components, collagens, for degradation; studies show that the aggrecanase Adamts4/5 are the principal enzymes responsible for degradation of the other principal component, aggrecan. Adamts5 is one of the shorter members of the zinc-dependent Adamts enzyme family[63,64] that has two thrombospondin (TS) motifs. Adamts4 is the shortest member of the zinc-dependent Adamts enzyme, containing only one TS motif. Full-length Adamts4 and Adamts5 are proenzymes that are activated by removing their prodomains via furin or furin-like enzymes.[65] Adamts4 and Adamts5 are two of the most active enzymes in aggrecan cleavage. Adamts4 is shown to be active during cartilage degeneration, and its expression is upregulated in degenerative cartilage.[66,67] Adamts5 has been shown to be active in both normal and degenerated cartilage.[68] It was not clear which of the Adamts family members is more important in the development of cartilage degenerative diseases until both *Adamts4* and *Adamts5* knockout mice were generated. Both *Adamts4* and *Adamts5* knockout mice are normal and have no gross abnormalities.[69–70] However, it was shown that *Adamts5* may play a more important role in OA development than *Adamts4*. Meniscal destabilization experiments were performed in both *Adamts4* and *Admats5* knockout mice showing that deletion of the *Adamts4* gene cannot protect OA progression, while deletion of the *Adamts5* gene alone decelerated cartilage degradation.[71,72]

An important aspect of defining the peptidases that are truly active in OA *in vivo* is the detection of specific cleavage products in the cartilage, synovial fluid, serum, or urine. Measurement of transcript abundance by quantitative PCR and/or peptidase abundance by immunoassay can generate correlative data, but they do not provide the definitive proof obtained with highly specific neo-epitope antibodies. Within this constraint, only a limited number of peptidases can be directly implicated in human OA cartilage pathology. These include one that cleaves in the C-telopeptide region of collagen II,[73] a bona fide collagenase; probably MMP13, which cleaves collagen II at Gly775-Leu/Ile776;[74,75] Cathepsin K, which degrades collagen II at Gly192-Lys193;[76] and an Adamts-aggrecanase, which cleaves aggrecan at Glu373-Ala374.[77] Whereas the expression and/or secretion of many other peptidases, such as MMP3,[78] MMP2, MMP9, and the PA/plasmin system,[79] are also increased in human OA, data to definitively establish *in vivo* activity are not yet available.

Further investigation of the mechanism by which *Adamts5* deletion protects cartilage showed that it may not due to elimination of aggrecanase activity from the cartilage.[26] This finding showed that Adamts5 is not, as previously concluded,[70,71] the major aggrecanase in mouse cartilage. However, a photographic, histologic, and biochemical examination of the "protected" joints in *Adamts5* knockout mice showed that the protection was, in fact, due to an elimination of fibrous overgrowth from the periarticular tissues and deposition of newly synthesized aggrecan in the cartilage.[26] This finding suggests that Adamts5 activity is profibrogenic in

cellular responses to joint injury and that deletion of the *Adamts5* gene switches cells to a chondrogenic phenotype. An explanation for these effects of *Adamts5* knockout on collagenous tissues was provided by an analysis of Smad-signaling pathways in wild-type and *Adamts5*-deficient fibroblasts and chondrocytes. This showed that in the presence of the enzyme, TGF-β1–mediated signaling is primarily through Smad2/3, leading to increased expression of fibrogenic genes such as type 1 and type III collagen. Conversely, in the absence of the enzyme and the presence of accumulated pericellular aggrecan, TGF-β1–mediated signaling is primarily through the Smad1/5/8 pathway. Activation of this pathway, which is also activated by BMP7 signaling, activates expression of chondrogenic genes such as aggrecan.[3] The precise mechanism by which a lack of Adamts5 activity promotes TGF-β1–mediated chondrogenesis is not known, but analysis of *Adamts5/CD44* double knockout mice shows that it is dependent on the presence of the hyaluronan receptor, CD44.[80] These data lead to the hypothesis that Adamts5 specifically degrades pericellular aggrecan in OA and that other aggrecanases, such as Adamts4, are responsible for degradation of the bulk of the tissue aggrecan, which resides in the intercellular matrix. If this model is supported by further work, it will directly affect strategies for therapeutic control of human OA.

Acknowledgments

This work was supported by Grants R01-AR055915 and R01-AR054465 to D.C. from the National Institutes of Health (NIH) and by Contract C024320 to D.C. from the New York State Department of Health and the Empire State Stem Cell Board.

Conflicts of interest

The authors declare no conflicts of interest.

References

1. Meulenbelt, I., M. Kloppenburg, H.M. Kroon, *et al.* 2007. Clusters of biochemical markers are associated with radiographic subtypes of osteoarthritis (OA) in subject with familial OA at multiple sites. The GARP study. *Osteoarthr. Cartilage* **15:** 379–385.
2. McGonagle, D., A.L. Tan, Carey, J. & M. Benjamin. 2010. The anatomical basis for a novel classification of osteoarthritis and allied disorders. *J. Anat.* **216:** 279–291.
3. Plaas, A., J. Velasco, D.J. Gorski, *et al.* 2011. The relationship between fibrogenic TGFb1 signaling in the joint and

cartilage degradation in post-injury osteoarthritis. *Osteoarthr. Cartilage* **19**: 1081–90.

4. Bijlsma, J.W., F. Berenbaum & F.P. Lafeber. 2010. Osteoarthritis: an update with relevance for clinical practice. *Lancet* **377**: 2115–2126.

5. Anderson, D.D., S. Chubinskaya, F. Guilak, *et al.* 2011. Post-traumatic osteoarthritis: improved understanding and opportunities for early intervention. *J. Orthop. Res.* **29**: 802–809.

6. van den Berg, W.B. 2010. Osteoarthritis year 2010 in review: pathomechanisms. *Osteoarthr. Cartilage* **19**: 338–341.

7. Schroeppel, J.P., J.D. Crist, H.C. Anderson & J. Wang. 2011. Molecular regulation of articular chondrocyte function and its significance in osteoarthritis. *Histol. Histopathol.* **26**: 377–394.

8. Ehrlich, M.G., A.L. Armstrong, B.V. Treadwell & H.J. Mankin. 1986. Degradative enzyme systems in cartilage. *Clin. Orthop. Relat. Res.* **213**: 62–68.

9. Eyre, D.R., M.A. Weis & J.J. Wu. 2006. Articular cartilage collagen: an irreplaceable framework? *Eur. Cell. Mater*. **12**: 57–63.

10. Wu, J.J., M.A. Weis, L.S. Kim & D.R. Eyre. 2011. Type III collagen, a fibril network modifier in articular cartilage. *J. Biol. Chem.* **285**: 18537–18544.

11. Schumacher, B.L., J.A. Block, T.M. Schmid, *et al.* 1994. A novel proteoglycan synthesized and secreted by chondrocytes of the superficial zone of articular cartilage. *Arch. Biochem. Biophys.* **311**: 144–152.

12. Wells, T., C. Davidson, M. Morgelin, J.L. Bird, *et al.* 2003. Age-related changes in the composition, the molecular stoichiometry and the stability of proteoglycan aggregates extracted from human articular cartilage. *Biochem J.* **370**: 69–79.

13. Plaas, A.H., L.A. West, S. Wong-Palms & F.R. Nelson. 1998. Glycosaminoglycan sulfation in human osteoarthritis. Disease-related alterations at the non-reducing termini of chondroitin and dermatan sulfate. *J. Biol. Chem.* **273**: 12642–12649.

14. Quinn, T.M., E.B. Hunziker & H.J. Hauselmann. 2005. Variation of cell and matrix morphologies in articular cartilage among locations in the adult human knee. *Osteoarthr. Cartilage* **13**: 672–678.

15. Han, Y. & V. Lefebvre. 2008. L-Sox5 and Sox6 drive expression of the aggrecan gene in cartilage by securing binding of Sox9 to a far-upstream enhancer. *Mol. Cell. Biol.* **28**: 4999–5013.

16. Williams, R., I.M. Khan, K. Richardson, *et al.* 2011. Identification and clonal characterisation of a progenitor cell sub-population in normal human articular cartilage. *P.LoS. One.* **5**: e13246.

17. Elvenes, J., G. Knutsen, O. Johansen, *et al.* 2009. Development of a new method to harvest chondroprogenitor cells from underneath cartilage defects in the knees. *J. Orthop. Sci.* **14**: 410–417.

18. Cawston, T.E. & D.A. Young. 2010. Proteinases involved in matrix turnover during cartilage and bone breakdown. *Cell. Tissue. Res.* **339**: 221–235.

19. Sandy, J.D., P.J. Neame, R.E. Boynton & C.R. Flannery. 1991. Catabolism of aggrecan in cartilage explants. Identification of a major cleavage site within the interglobular domain. *J. Biol. Chem.* **266**: 8683–8685.

20. Ilic, M.Z., C.J. Handley, H.C. Robinson & M.T. Mok. 1992. Mechanism of catabolism of aggrecan by articular cartilage. *Arch. Biochem. Biophys.* **294**: 115–122.

21. Loulakis, P., A. Shrikhande, G. Davis & C.A. Maniglia. 1992. *N*-terminal sequence of proteoglycan fragments isolated from medium of interleukin-1-treated articular-cartilage cultures. Putative site(s) of enzymic cleavage. *Biochem. J.* **284**: 589–593.

22. Brew, C.J., P.D. Clegg, R.P. Boot-Handford, *et al.* 2011. Gene expression in human chondrocytes in late osteoarthritis is changed in both fibrillated and intact cartilage without evidence of generalised chondrocyte hypertrophy. *Ann. Rheum. Dis.* **69**: 234–240.

23. Aigner, T., K. Fundel, J. Saas, P.M. Gebhard, *et al.* 2006. Large-scale gene expression profiling reveals major pathogenetic pathways of cartilage degeneration in osteoarthritis. *Arthritis Rheum.* **54**: 3533–3544.

24. Wei, T., N.H. Kulkarni, Q.Q. Zeng, *et al.* 2010. Analysis of early changes in the articular cartilage transcritisome in the rat meniscal tear model of osteoarthritis: pathway comparisons with the rat anterior cruciate transection model and with human osteoarthritic cartilage. *Osteoarthr. Cartilage* **18**: 992–1000.

25. Little, C.B., A. Barai, D. Burkhardt, *et al.* 2009. Matrix metalloproteinase 13-deficient mice are resistant to osteoarthritic cartilage erosion but not chondrocyte hypertrophy or osteophyte development. *Arthritis Rheum.* **60**: 3723–3733.

26. Li, J., W. Anemaet, M.A. Diaz, *et al.* 2011. Knockout of ADAMTS5 does not eliminate cartilage aggrecanase activity but abrogates joint fibrosis and promotes cartilage aggrecan deposition in murine osteoarthritis models. *J. Orthop. Res.* **4**: 516–522.

27. Moodie, J.P., Stok, K.S., R. Muller, *et al* . 2011. Multimodal imaging demonstrates concomitant changes in bone and cartilage after destabilisation of the medial meniscus and increased joint laxity. *Osteoarthr. Cartilage* **19**: 163–170.

28. Blaney Davidson, E.N., P.M. van der Kraan & W.B. van den Berg. 2007. TGF-β and osteoarthritis. *Osteoarthr. Cartilage* **15**: 597–604.

29. Serra, R., M. Johnson, E.H. Filvaroff, *et al.* 1997. Expression of a truncated, kinase-defective TGF-β type II receptor in mouse skeletal tissue promotes terminal chondrocyte differentiation and osteoarthritis. *J. Cell Biol.* **139**: 541–552.

30. Yang, X., L. Chen, X. Xu, *et al.* 2001. TGF-β Smad3 signals repress chondrocyte hypertrophic differentiation and are required for maintaining articular cartilage. *J. Cell Biol.* **153**: 35–46.

31. Zuscik, M.J., J.F. Baden, Q. Wu, *et al.* 2004. 5-Azacytidine alters TGF-β and BMP signaling and induces maturation in articular chondrocytes. *J. Biol. Chem.* **92**: 316–331.

32. Wu, Q., M. Wang, M.J. Zuscik, *et al.* 2008. Regulation of embryonic endochondral ossification by Smurf2. *J. Orthop. Res.* **26**: 704–712.

33. Wu, Q., K.O. Kim, E.R. Sampson, *et al.* 2008. Smurf2 induces an osteoarthritis-like phenotype and degrades phosphorylated Smad3 in transgenic mice. *Arthritis Rheum.* **58**: 3132–3144.

34. Valdes, A.M., T.D. Spector, A. Tamm, *et al.* 2010. Genetic variation in the *Smad3* gene is associated with hip and knee osteoarthritis. *Arthritis Rheum.* **62:** 2347–2352.

35. Yang, Y. 2003. Wnts and Wing: Wnt signaling in vertebrate limb development and musculoskeletal morphogenesis. *Birth. Defects. Res. (PartC)* **69:** 305–317.

36. Zhu, M., D. Tang, Q. Wu, *et al.* 2009. Activation of β-catenin signaling in articular chondrocytes leads to osteoarthritis-like phenotype in adult β-catenin conditional activation mice. *J. Bone. Miner. Res.* **24:** 12–21.

37. Clevers, H. 2006. Wnt/β-catenin signaling in development and disease. *Cell* **127:** 469–80.

38. Day, T.F., X. Guo, L. Garrett-Beal & Y. Yang. 2005. Wnt/β-catenin signaling in mesenchymal progenitors controls osteoblast and chondrocyte differentiation during vertebrate skeletogenesis. *Dev. Cell.* **8:** 739–750.

39. Loughlin, J., B. Dowling, K. Chapman, *et al.* 2004. Functional variants within the secreted frizzled-related protein3 gene are associated with hip OA in females. *Proc. Natl. Acad. Sci. USA.* **101:** 9757–9762.

40. Spector, T.D. & A.J. MacGregor. 2004. Risk factors for osteoarthritis: genetics [review]. *Osteoarthr. Cartilage* **12** (Suppl A): S39–S44.

41. Slagboom, P.E., B.T. Heijmans, M. Beekman, *et al.* 2000. Genetics of human aging: the search for genes contributing to human longevity and diseases of the old [review]. *Ann. N.Y. Acad. Sci.* **908:** 50–63.

42. Loughlin, J., Z. Mustafa, A. Smith, *et al.* 2000. Linkage analysis of chromosome 2q in osteoarthritis. *Rheumatology* **39:** 377–381.

43. Enomoto-Iwamoto, M., J. Kitagaki, E. Koyama, *et al.* 2002. The Wnt antagonist Frzb-1 regulates chondrocyte maturation and long bone development during limb skeletogenesis. *Dev. Biol.* **251:** 142–156.

44. Tamamura, Y., T. Otani, N. Kanatani, *et al.* 2005. Developmental regulation of Wnt/β-catenin signals is required for growth plate assembly, cartilage integrity, and endochondral ossification. *J. Biol. Chem.* **280:** 19185–19195.

45. Lories, R.J., J. Peeters, A. Bakker, *et al.* 2007. Articular cartilage and biomechanical properties of the long bones in *Frzb*-knockout mice. *Arthritis Rheum.* **56:** 4095–4103.

46. Blom, A.B., S.M. Brockbank, P.L. van Lent, *et al.* 2009. Involvement of the Wnt signaling pathway in experimental and human osteoarthritis: prominent role of Wnt-induced signaling protein 1. *Arthritis Rheum.* **60:** 501–512.

47. Chun, J.S., H. Oh, S. Yang & M. Park. 2008. Wnt signaling in cartilage development and degeneration. *BMB. Rep.* **41:** 485–494.

48. Tian, H., S.L. McKnight & D.W. Russell. 1997. Endothelial PAS domain protein 1 (EPAS1), a transcription factor selectively expressed in endothelial cells. *Genes. Dev.* **11:** 72–82.

49. Patel, S.A. & M.C. Simon. 2008. Biology of hypoxia-inducible factor-2α in development and disease. *Cell Death Differ.* **15:** 628–634.

50. Tian, H., R.E. Hammer, A.M. Matsumoto, *et al.* 1998. The hypoxia-responsive transcription factor EPAS1 is essential for catecholamine homeostasis and protection against heart failure during embryonic development. *Genes. Dev.* **12:** 3320–3324.

51. Saito, T., A. Fukai, A. Mabuchi, *et al.* 2010. Transcriptional regulation of endochondral ossification by HIF-2α during skeletal growth and osteoarthritis development. *Nat. Med.* **16:** 678–686.

52. Yang, S., J. Kim, J.H. Ryu, *et al.* 2010. Hypoxia-inducible factor-2α is a catabolic regulator of osteoarthritic cartilage destruction. *Nat. Med.* **16:** 687–693.

53. Lin, A.C., B.L. Seeto, J.M. Bartoszko, *et al.* 2009. Modulating hedgehog signaling can attenuate the severity of osteoarthritis. *Nat. Med.* **15:** 1421–1426.

54. Mak, K.K., M.H. Chen, T.F. Day, *et al.* 2006. Wnt/β-catenin signaling interacts differentially with Ihh signaling in controlling endochondral bone and synovial joint formation. *Development* **133:** 3695–3707.

55. Vincenti, M.P. & C.E. Brinckerhoff. 2002. Transcriptional regulation of collagenase (MMP-1, MMP-13) genes in arthritis: integration of complex signaling pathways for the recruitment of gene-specific transcription factors. *Arthritis Res.* **4:** 157–164.

56. Shiomi, T., V. Lemaître, J. D'Armiento & Y. Okada. 2010. Matrix metalloproteinases, a disintegrin and metalloproteinases, and a disintegrin and metalloproteinases with thrombospondin motifs in non-neoplastic diseases. *Pathol. Int.* **60:** 477–496.

57. Knäuper, V., C. Lopez-Otin, B. Smith, G. Knight & G. Murphy. 1996. Biochemical characterization of human collagenase-3. *J. Biol. Chem.* **271:** 1544–1550.

58. Reboul, P., J.P. Pelletier, G. Tardif, *et al.* 1996. The new collagenase, collagenase-3, is expressed and synthesized by human chondrocytes but not by synoviocytes: a role in osteoarthritis. *J. Clin. Invest.* **97:** 2011–2019.

59. Roach, H.I., N. Yamada, K.S. Cheung, *et al.* 2005. Association between the abnormal expression of matrix-degrading enzymes by human osteoarthritic chondrocytes and demethylation of specific CpG sites in the promoter regions. *Arthritis Rheum.* **52:** 3110–3124.

60. Stickens, D., D.J. Behonick, N. Ortega, *et al.* 2004. Altered endochondral bone development in matrix metalloproteinase 13-deficient mice. *Development* **131:** 5883–5895.

61. Inada, M., Y. Wang, M.H. Byrne, *et al.* 2004. Critical roles for collagenase-3 (Mmp13) in development of growth plate cartilage and in endochondral ossification. *Proc. Natl. Acad. Sci. USA.* **101:** 17192–17197.

62. Neuhold, L.A., L. Killar, W. Zhao, *et al.* 2001. Postnatal expression in hyaline cartilage of constitutively active human collagenase-3 (MMP-13) induces osteoarthritis in mice. *J. Clin. Invest.* **107:** 35–44.

63. Gomis-Rüth, F.X. 2003. Structural aspects of the metzincin clan of metalloendopeptidases. *Mol. Biotechnol.* **24:** 157–202.

64. Tang, B.L. 2001. ADAMTS: a novel family of extracellular matrix proteases. *Int. J. Biochem. Cell. Biol.* **33:** 33–44.

65. Fosang, A.J. & C.B. Little. 2008. Drug insight: aggrecanases as therapeutic targets for osteoarthritis. *Nat. Clin. Pract. Rheumatol.* **4:** 420–427.

66. Song, R.H., M.D. Tortorella, A.M. Malfait, *et al.* 2007. Aggrecan degradation in human articular cartilage explants is mediated by both ADAMTS-4 and ADAMTS-5. *Arthritis Rheum.* **56:** 575–585.

67. Tortorella, M.D., R.Q. Liu, T. Burn, *et al.* 2002. Characterization of human aggrecanase 2 (ADAM-TS5): substrate specificity studies and comparison with aggrecanase 1 (ADAM-TS4). *Matrix Biol.* **21:** 499–511.

68. Kevorkian, L., D.A. Young, C. Darrah, *et al.* 2004. Expression profiling of metalloproteinases and their inhibitors in cartilage. *Arthritis Rheum.* **50:** 131–141.

69. Glasson, S.S., R. Askew, B. Sheppard, *et al.* 2004. Characterization of and osteoarthritis susceptibility in ADAMTS-4-knockout mice. *Arthritis Rheum.* **50:** 2547–2558.

70. Glasson, S.S., R. Askew, B. Sheppard, *et al.* 2005. Deletion of active ADAMTS5 prevents cartilage degradation in a murine model of osteoarthritis. *Nature* **434:** 644–648.

71. Stanton, H., F.M. Rogerson, C. J. East, *et al.* 2005. ADAMTS5 is the major aggrecanase in mouse cartilage in vivo and in vitro. *Nature* **434:** 648–652.

72. Echtermeyer, F., J. Bertrand, R. Dreier, *et al.* 2009. Syndecan-4 regulates ADAMTS-5 activation and cartilage breakdown in osteoarthritis. *Nat. Med.* **15:** 1072–1076.

73. Lohmander, L.S., L.M. Atley, T.A. Pietka & D.R. Eyre. 2003. The release of crosslinked peptides from type II collagen into human synovial fluid is increased soon after joint injury and in osteoarthritis. *Arthritis Rheum.* **48:** 3130–3139.

74. Poole, A.R., M. Ionescu, M.A. Fitzcharles & R.C. Billinghurst. 2004. The assessment of cartilage degradation in vivo: development of an immunoassay for the measurement in body fluids of type II collagen cleaved by collagenases. *J. Immunol. Methods* **294:** 145–153.

75. Charni, N., F. Juillet & P. Garnero. 2005. Urinary type II collagen helical peptide (HELIX-II) as a new biochemical marker of cartilage degradation in patients with osteoarthritis and rheumatoid arthritis. *Arthritis Rheum.* **52:** 1081–1090.

76. Dejica, V.M., J.S. Mort, S. Laverty, *et al.* 2008. Cleavage of type II collagen by cathepsin K in human osteoarthritic cartilage. *Am. J. Pathol.* **173:** 161–169.

77. Sandy, J.D., C.R. Flannery, P.J. Neame & L. S. Lohmander. 1992. The structure of aggrecan fragments in human synovial fluid. Evidence for the involvement in osteoarthritis of a novel proteinase which cleaves the Glu 373-Ala 374 bond of the interglobular domain. *J. Clin. Invest.* **89:** 1512–1516.

78. Lohmander, L.S., L. Dahlberg, D. Eyre, *et al.* 1998. Longitudinal and cross-sectional variability in markers of joint metabolism in patients with knee pain and articular cartilage abnormalities. *Osteoarthr. Cartilage* **6:** 351–361.

79. Yang, S.F., Y.S. Hsieh, K.H. Lue, *et al.* 2008. Effects of nonsteroidal anti-inflammatory drugs on the expression of urokinase plasminogen activator and inhibitor and gelatinases in the early osteoarthritic knee of humans. *Clin. Biochem.* **41:** 109–116.

80. Velasco, J., J. Li, L. Dipietro, *et al.* 2011. Adamts5 deletion blocks murine dermal repair through CD44-mediated aggrecan accumulation and modulation of transforming growth factor β1 (TGF-β) signaling. *J. Biol. Chem.* **286:** 26016–26027.

Ann. N.Y. Acad. Sci. ISSN 0077-8923

ANNALS OF THE NEW YORK ACADEMY OF SCIENCES
Issue: *Skeletal Biology and Medicine II*

Pathogenesis of diabetic neuropathy: bad to the bone

Lawrence Chan,[1] Tomoya Terashima,[1,2] Hiroshi Urabe,[1] Fan Lin,[1] and Hideto Kojima[1,2]

[1]Diabetes and Endocrinology Research Center, Baylor College of Medicine, 1 Baylor Plaza, Houston, Texas 77030.
[2]Department of Molecular Genetics in Medicine, Shiga University of Medical Science, Tsukinowacho, Seta, Otsu, Shiga 520-2192, Japan

Address for correspondence: Dr. Lawrence Chan, Diabetes & Endocrinology Research Center, Baylor College of Medicine, 1 Baylor Plaza, Houston, TX 77030. lchan@bcm.edu

Insulin and proinsulin are normally produced only by the pancreas and thymus. We detected in diabetic rodents the presence of extra pancreatic proinsulin-producing bone marrow-derived cells (PI-BMDCs) in the BM, liver, and fat. In mice and rats with diabetic neuropathy, we also found proinsulin-producing cells in the sciatic nerve and neurons of the dorsal root ganglion (DRG). BM transplantation experiments using genetically marked donor and recipient mice showed that the proinsulin-producing cells in the DRG, which morphologically resemble neurons, are actually polyploid proinsulin-producing fusion cells formed between neurons and PI-BMDCs. Additional experiments indicate that diabetic neuropathy is not simply the result of nerve cells being damaged directly by hyperglycemia. Rather, hyperglycemia induces fusogenic PI-BMDCs that travel to the peripheral nervous system, where they fuse with Schwann cells and DRG neurons, causing neuronal dysfunction and death, the *sine qua non* for diabetic neuropathy. Poorly controlled diabetes is indeed bad to the bone.

Keywords: diabetic complications; proinsulin; bone marrow-derived cells; cell fusion

Introduction

The worldwide diabetes epidemic has shown no signs of abatement. Its prevalence has more than doubled since 1980, increasing from an estimated 153 million, three decades ago to about 347 million in 2008.[1] Diabetic people suffer much morbidity and premature mortality because of chronic diabetic complications, which include cardiovascular and cerebrovascular disease, nephropathy, retinopathy, and neuropathy. Multiple pathogenic effectors downstream of hyperglycemia contribute to chronic diabetic complications.[2–14] Diabetic neuropathy is the most common diabetic complication, afflicting over 50% of all diabetics, and is the leading cause of nontraumatic limb amputations. The biochemical perturbations that underlie diabetic neuropathy are very similar to those of other complications; they include oxidative stress,[4,15] activation of the polyol pathway,[7] increased advanced glycation end products and their receptors,[15] activation of protein kinase C (PKC)[9] and mitogen-activated pro-

tein kinases (MAPK),[10] and inducible nitric oxide synthase.[16] Furthermore, hypoxia and ischemia,[6] elevated cytokines, such as tumor necrosis factor (TNF)-α[17] and IL-6,[18] and nerve growth factor (NGF) deficiency[19] also play important etiologic roles in diabetic neuropathy.

The consensus in the field is that in diabetes these dysregulated metabolic pathways produce malfunction, injury, and death of cells that are intrinsic to the peripheral nervous system, that is, Schwann cells, vasa nervorum, and dorsal root ganglia (DRG) neurons. Recent studies in our laboratory have revealed that a specific cell type that is extrinsic to the nervous system seems to also play a causative role in diabetic neuropathy. These diabetes-specific cells are a subpopulation of circulating bone marrow (BM)-derived cells that travel to the peripheral nervous system where they produce major tissue damage by a novel mechanism.

A few years ago, when we were working on a gene therapy for diabetes in streptozotocin (STZ)-induced diabetic mice, we made a surprising

doi: 10.1111/j.1749-6632.2011.06309.x

Ann. N.Y. Acad. Sci. 1240 (2011) 70–76 © 2011 New York Academy of Sciences.

observation. Our new gene therapy formulation reversed diabetes by inducing periportal insulin-producing beta-like cells in the liver that exhibit robust glucose-stimulated insulin secretion.[20] Insulin transcripts were undetectable in the liver of nondiabetic mice. Unexpectedly, we detected tiny amounts of insulin transcripts in the liver of untreated and empty vector-treated STZ-diabetic mice.[20] Subsequently, we discovered that diabetes leads to the appearance of proinsulin (PI)-producing cells in multiple organs and tissues in rodents. These abnormal cells are present in STZ-induced insulin-deficient diabetes and in obese animals that have developed hyperinsulinemia and type 2 diabetes induced by high fat-diet feeding. These unique cells also occur in *ob/ob* mice, which have low level hyperglycemia for months before analysis.[21] Interestingly, the same cells also express the proinflammatory cytokine TNF-α.[22] Further examination revealed that these cells first show up in the BM, when they travel to different peripheral organs and tissues by the circulation. The proinsulin-producing bone marrow-derived cells (PI-BMDCs) appear within one to three days of intermittent hyperglycemia induced by repeated glucose injections. Our initial conjecture was that one could use such PI-BMDCs for diabetes cell therapy. Shortly after our discovery, Oh *et al.* reported that incubation in high-glucose medium induces insulin-gene transcription in BM cells *in vitro*.[23] They transferred these cells to the kidney capsule of diabetic rodents and reported that the maneuver partially ameliorated the hyperglycemia, although the authors did not report the serum insulin level. When we investigated the possible utility of these cells for diabetes cell therapy, we found that the tiny amounts of PI produced by the PI-BMDCs was not detectably secreted into the culture medium, and thus, PI-BMDCs induced by exposure to a high-glucose medium have no therapeutic potential for diabetes-cell therapy.

The PI-BMDCs are F4/80[+] cells, and morphological examination indicates that they are mostly macrophage-like cells. We used immunohistochemistry to localize PI-producing cells in other tissues of rodents and found PI[+] in relative abundance in the peripheral nervous system of long-term diabetic mice and rats. By light microscopy, the PI[+] cells appear like normal nerve cells; PI[+] signals are readily detected in sciatic nerves and neurons in the DRG, but are seen only in rodents that have diabetes for

over a month, when the animals have developed impaired nerve function, that is, the PI-producing neurons appear at the same time that the animals develop diabetic neuropathy. We incubated isolated DRG neurons in high-glucose medium *in vitro* and found that, unlike BM cells, neurons exposed to a high-glucose medium fail to turn on PI production. We hypothesized that these PI-producing neuron-like cells represent PI-BMD–neuron fusion cells.

The fusion of BMDCs with cells in different tissues and organs, for example, skeletal muscle, cardiomyocytes, hepatocytes, brain, and intestine, has been reported by different laboratories.[24–29] The fusion of hematopoietic cells with Purkinje neurons has been found to occur in rodents and humans;[24,30,31] the frequency of these fusion cells was increased in the presence of inflammation and after X-radiation.[32,33] These were tantalizing reports, but the pathophysiological significance of fusion cells in the central nervous system remained unclear. These reports of the occurrence of fusion of BMDCs and Purkinje cells prompted us to reexamine the PI[+] nerve cells that appear in rodents with diabetic neuropathy in a new light. We reasoned that the diabetes-induced PI-producing neurons might represent fusion cells formed between neurons and PI-BMDCs, with the capacity for PI production coming from the BMDC fusion partner.

PI-BMDCs in multiple tissues of diabetic rodents

We observed the presence of immunoreactive PI[+] cells in both paraffin sections and frozen liver sections of diabetic mice (Fig. 1A).[21] Further, PI[+] cells were also observed in histological sections from adipose tissue and BM of diabetic mice and rats. The presence of insulin mRNA in the liver was corroborated by *in situ* nucleic acid hybridization.[21] The co-occurrence of insulin transcripts and immunoreactive PI indicates that the transcripts were translated into protein. To confirm that insulin-gene transcription occurs in the same cells that contain immunoreactive PI, we induced STZ-diabetes in mouse insulin promoter-green fluorescent protein (MIP-GFP) transgenic mice, which express GFP driven by the insulin promoter,[34] and found that diabetes led to the expression of GFP in the same cells that harbor immunoreactive PI (Fig. 1B). Immunohistochemical analysis of the liver, fat, and BM of *ob/ob* mice as well as high fat-diet fed mice revealed

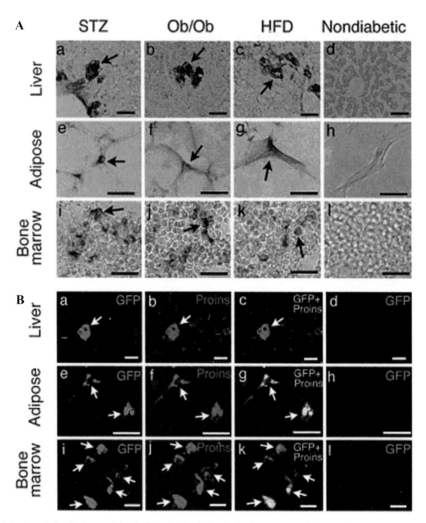

Figure 1. (A) Proinsulin+ cells (arrows) in the liver (a–d), abdominal adipose tissue (e–h), and bone marrow (i–l) of STZ (a, e, and i), ob/ob (b, f, and j), and high-fat diet (HFD; c, g, and k) diabetic mice and nondiabetic mice (d, h, and l). In nondiabetic mice, proinsulin+ cells were not found in the liver (d), adipose tissue (h), or bone marrow (l). (Scale bars, 25 μm.) (B) Overlap images of GFP/proinsulin in the liver (a–d), adipose tissue (e–h), and bone marrow (i–l) from STZ-induced diabetic (a–c, e–g, and i–k) and nondiabetic (d, h, and l) MIP-GFP mice. GFP and proinsulin signals completely overlap (arrows). (Scale bars: 25 μm, a–d; 20 μm, e–h; 10 μm, i–l). From Ref. 21 (with permission).

PI+ cells also occur in these type 2 diabetes models (Fig. 1A).[21]

PI-BMDCs fuse with DRG neurons in mice with diabetic neuropathy

We produced diabetes by STZ treatment in eight-week-old rats and mice. Eight to twelve weeks later, all animals had developed diabetic neuropathy as evidenced by abnormal nerve function tests (impaired motor nerve conduction velocity, compound muscle action potential, sensory nerve conduction velocity, and sensory nerve action potential). By immunohistochemistry, we detected PI+ cells in the DRG and sciatic nerve of diabetic rats but not nondiabetic animals. Approximately, 10% of DRG neurons expressed PI 12 weeks after STZ-induced diabetes. The presence of insulin transcripts in DRG neurons was corroborated by *in situ* hybridization.[35]

We performed a bone marrow transplantation (BMT) experiment, transferring BM cells from

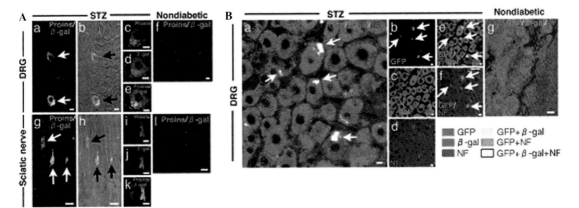

Figure 2. (A) Wild-type mice received BMT from β-gal donors. Immunofluorescence double staining of Proins/β-gal of DRG and sciatic nerves at eight weeks after STZ (a–e, g–k) or in nondiabetic recipients (f and l). Bright images are shown for comparison. (B) BMT of MIP-GFP donors[34] to β-gal recipients. Sections taken from DRG eight weeks after STZ-induced diabetes in half the recipients. Immunofluorescence triple staining of GFP/β-gal/NF. From Ref. 35 (with permission).

transgenic mice that constitutively expressed β-gal to wild-type C57BL/6 mice, and used STZ to induce diabetes in half of the BMT recipients. By immunostaining, we found PI⁺ cells in the DRG and sciatic nerve (Fig. 2A) of diabetic mice; these cells displayed overlapping PI and β-gal immunostaining, indicating that they originated from the BM. Neither PI nor β-gal was detected in sections from nondiabetic BMT recipients (Fig. 2A). These observations were corroborated by data obtained in β-gal transgenic mice that have received BMT from MIP-GFP transgenic mice,[34] which express GFP driven by the mouse insulin promoter (Fig. 2B).[35] Additional experiments indicate that in the peripheral nervous system of diabetic animals, some of the fusion cells coexpress PI and neurofilament, a neuron-specific protein, and others coexpress PI and S100, a Schwann cell protein, indicating that both neurons and Schwann cells are individually involved in fusing with PI-BMDCs (see Fig. 1D in Terashima *et al.*[35]). Finally, when we performed BMT from male BM donors to female recipients, we detected Y chromosome-positive fusion cells in the DRG neurons of recipient mice with diabetic neuropathy,[36] lending further support for a BM-neuron cell fusion event.

To estimate the frequency of PI-BMDC–neuron fusion, we analyzed individual neurons isolated from the DRG of nondiabetic and diabetic animals and found that in nondiabetic rats,[35] 99.1% of the DRG neurons were diploid (2n); tetraploid (4n) cells constituted only 0.9%, and no cells were higher

than tetraploid. In contrast, among neurons isolated from the DRG of diabetic rats, a substantially lower proportion, 86.6%, were diploid (2n), and there was a much higher proportion of polyploid cells (i.e., 12.5% were tetraploid, and 0.9% contained 6n or 8n [0.7% 6n and 0.2% 8n]).[35] When we analyzed the distribution of PI immunostaining among the isolated DRG neurons in diabetic rats, we found that polyploidy occurs exclusively in PI-expressing cells, further indicating that PI expression is a marker for fusion cells. Findings very similar to those in diabetic rats have been reported recently in mice with diabetic neuropathy.[36]

Fusion of PI-BMDCs and DRG neurons *in vitro*

We incubated DRG neurons from Rose26-flox-stop-GFP transgenic mice (which have a loxP-flanked STOP sequence in front of the green fluorescent protein gene knocked into the Rosa26 locus in all tissues) with RIP-Cre BMDCs (from transgenic mice that express Cre recombinase driven by the rat insulin promoter) in the presence of high- and low-glucose medium, and found PI-BMDC–neuron fusion cell formation *in vitro* only in the presence of high glucose, but not low glucose medium. These fusion cells were marked by GFP expression. When the PI-BMDCs fuse with the neurons, the high glucose-induced Cre recombinase produced by the RIP-Cre BMDCs will access and delete the STOP signal in front of the GFP in the Rosa26-floz-stop-GFP neuron fusion partner, allowing the GFP to be expressed in the fusion cell. Therefore, the role of a high-glucose environment in the induction of PI-BMDCs and in facilitating PI-BMDC–neuron cell fusion can be reproduced under *in vitro* incubation conditions.[36]

Discussion

With the advances in glucose monitoring and individualized insulin therapy in the last two decades, it is now rare for diabetics to die of ketoacidosis. The much longer life span of diabetics in the 21st century translates into many more patients who suffer from chronic diabetic complications, including a large proportion of patients who die from them. Hyperglycemia is known to underlie most, if not all, diabetic complications. According to one scenario:

> Intracellular hyperglycemia . . . causes increased mitochondrial production of ROS (reactive oxygen species). The ROS causes strand breaks in nuclear DNA, which activate PARP (poly[ADP-ribose] polymerase). PARP then modifies GAPDH (*glyceraldehydes-3 phosphate dehydrogenase*), thereby reducing its activity. Finally, decreased GADPH activates the polyol pathway, increases intracellular AGE (*advanced glycation end product*) formation, activates PKC (*protein kinase C*) and subsequently NF-κB, and activates hexosamine pathway flux" and "subsequent modification of proteins by *N*-acetylglucosamine" [notations in italics added by authors].[37]

Similar mechanisms, with minor variations in emphasis, have been proposed by others.[2–14] Most of these mechanisms have also been documented to operate in diabetic neuropathy.[38–40] In all cases, the culprit cells that underlie the neuropathy are assumed to be local cells in the different target organ system affected by the particular complication.

Different groups have described the fusion of hematopoietic cells with Purkinje neurons in rodents and humans.[24,30,31] Hematopoietic cells derived from the BM are often thought to function as "healers," playing a positive role in neural regeneration in neuropathy. Despite the absence of functional experiments, there is widespread speculation that BMDC–Purkinje fusion represents a mechanism that the body uses to repair damaged tissues.[32,33,41–43] Experiments in our laboratory showed that PI-BMDC–neuron fusion is far from being a beneficial healing event. PI-BMDCs are a subpopulation of highly proinflammatory BMDCs specifically induced by diabetes that have fusogenic

properties. They seem to play a central role in the pathogenesis of diabetic neuropathy.

Hyperglycemia induces ROS, which in turn activates poly(ADP-ribose) polymerase (PARP), which plays a central role in the pathogenesis of multiple diabetic complications. Neuronal PARP activation has been shown to occur in the spinal cord, peripheral nerves, and DRG of rodents with diabetic neuropathy. PARP inhibition was found to protect against diabetic neuropathy.[39,44–50] Furthermore, PARP-knockout mice with a mixed genetic background[51] or C57BL/6 background[36] are resistant to diabetic neuropathy. The dogma among researchers that PARP activation within peripheral nerves, including Schwann cells, vasa nervorum, and DRG neurons is "an early and fundamental mechanism of peripheral diabetic neuropathy."[52] We note that major inflammatory pathways, for example, PKC and NF-κB, are directly downstream of PARP and much of the inflammation in diabetic complication is mediated by proinflammatory hematopoietic cells. PI-BMDCs are a diabetes-specific cell type that mediates much of the inflammation-related pathology of diabetic complications. Importantly, these same cells have great fusogenic potential that can be demonstrated under *in vitro* conditions.[36] They directly cause tissue damage by fusing with both neurons and Schwann cells and are thus a major culprit in diabetic neuropathy.[35] To further dissect the role of PARP in diabetic neuropathy, we recently genetically inactivated PARP globally, and in a BM-specific manner.[36] We found that BMT of PARP-knockout BM cells to wild-type mice protects against, and conversely, BMT of wild-type cells to PARP-knockout mice confers susceptibility to, diabetic neuropathy. In other words, the major determinant in diabetic neuropathy in the BMT model is the PARP genotype of the BM cells. *In vitro* incubation experiments further indicate that in high-glucose PARP$^+$ (wild-type) PI-BMDCs are fusogenic. Loss of PARP in BMDCs makes them resistant to high-glucose–induced PI expression; it also makes them lose most of their fusogenic potential despite the presence of high glucose.[36]

In summary, the hyperglycemia in diabetes induces the formation of fusogenic PI-BMDCs, proinflammatory cells that coexpress TNF-α. The high-glucose environment further facilitates the fusion of these cells with DRG neurons, causing nerve dysfunction and premature apoptosis, events that lead to diabetic neuropathy. Over the last decade, research in the area of skeletal biology has revealed that, in addition to controlling calcium homeostasis, skeletal cells and secretions also play key roles in energy and glucose metabolism.[53] Conversely, here we show that diabetes causes BM cells to misbehave and become promiscuous, wreaking havoc as they travel to different tissues and organs by fusing with local cells. In addition to fusing with neurons and Schwann cells, we showed that PI-BMDCs also fuse with hepatocytes in diabetic animals.[22] Interestingly, the pathological consequence of PI-BMDC–hepatocyte fusion in the liver is not readily apparent because of the large functional reserve of the liver. To date, diabetic neuropathy is the only chronic diabetic complication proven to be a direct victim of PI-BMDCs' misadventures. Given the ruinous potential of this unique cell population in diabetic animals, we will be surprised if these messengers of destruction are not directly involved in other diabetic complications. Poorly controlled diabetes is bad to the bone.

Acknowledgments

This study was supported by NIH Grants HL-51586 (to LC) and P30DK079638 for the Diabetes and Endocrinology Research Center at Baylor College of Medicine. L.C. was also supported by the Betty Rutherford Chair for Diabetes Research from St. Luke's Episcopal Hospital (Houston, Texas) and the T.T. and W.F. Chao Global Foundation.

Conflicts of interest

The authors declare no conflicts of interest.

References

1. Danaei, G., M.M. Finucane, Y. Lu, *et al.* 2011. National, regional, and global trends in fasting plasma glucose and diabetes prevalence since 1980: systematic analysis of health examination surveys and epidemiological studies with 370 country-years and 2.7 million participants. *Lancet* **378:** 31–40.
2. Srinivasan, S., M. Stevens & J.W. Wiley. 2000. Diabetic peripheral neuropathy. Evidence for apptosis and associated mitochondrial dysfunction. *Diabetes* **49:** 1932–1938.
3. Schmeichel, A.M., J.D. Schmelzer & P.A. Low. 2003. Oxidative injury and apoptosis of dorsal root ganglion neurons in chronic experimental diabetic neuropathy. *Diabetes* **52:** 165–171.
4. Vincent, A.M., J.W. Russell, P. Low & E.L. Feldman. 2004. Oxidative stress in the pathogenesis of diabetic neuropathy. *Endocr. Rev.* **25:** 612–628.

5. Leininger, G.M., J.L. Edwards, M.J. Lipshaw & E.L. Feldman. 2006. Mechanisms of disease: mitochondria as new therapeutic targets in diabetic neuropathy. *Nat. Clin. Pract. Neurol.* **2:** 620–628.
6. Low, P.A., T.D. Lagerlund & P.G. McManis. 1989. Nerve blood flow and oxygen delivery in normal, diabetic, and ischemic neuropathy. *Int. Rev. Neurobiol.* **31:** 355–438.
7. Oates, P.J. 2002. Polyol pathway and diabetic peripheral neuropathy. *Int. Rev. Neurobiol.* **50:** 325–392.
8. Thornalley, P.J. 2002. Glycation in diabetic neuropathy: characteristics, consequences, causes, and therapeutic options. *Int. Rev. Neurobiol.* **50:** 37–57.
9. Xia, P., R.M. Kramer & G.L. King. 1995. Identification of the mechanism for the inhibition of Na+,K(+)-adenosine triphosphatase by hyperglycemia involving activation of protein kinase C and cytosolic phospholipase A2. *J. Clin. Invest.* **96:** 733–740.
10. Tomlinson, D.R. 1999. Mitogen-actiavated protein kinases as glucose transducers for diabetic complications. *Diabetologia* **42:** 1271–1281.
11. Russell, J.W., D. Golovoy, A.M. Vincent, *et al.* 2002. High glucose-induced oxidative stress and mitochondrial dysfunction in neurons. *FASEB J.* **16:** 1738–1748.
12. King, G.L. & M.R. Loeken. 2004. Hyperglycemia-induced oxidative stress in diabetic complications. *Histochem. Cell Biol.* **122:** 333–338.
13. Vincent, A.M., M. Brownlee & J.W. Russell. 2002. Oxidative stress and programmed cell death in diabetic neuropathy. *Ann. N.Y. Acad. Sci.* **959:** 368–383.
14. Groop, P.H., C. Forsblom & M.C. Thomas. 2005. Mechanisms of disease: pathway-selective insulin resistance and microvascular complications of diabetes. *Nat. Clin. Pract. Endocrinol. Metab.* **1:** 100–110.
15. Toth, C., L.L. Rong, C. Yang, *et al.* 2008. Receptor for advanced glycation end products (RAGEs) and experimental diabetic neuropathy. *Diabetes* **57:** 1002–1017.
16. Vareniuk, I., I.A. Pavlov & I.G. Obrosova. 2008. Inducible nitric oxide synthase gene deficiency counteracts multiple manifestations of peripheral neuropathy in a streptozotocin-induced mouse model of diabetes. *Diabetologia* **51:** 2126–2133.
17. Yamakawa, I., H. Kojima, T. Terashima, *et al.* 2011. Inactivation of TNFá ameliorates diabetic neuropathy in mice. *Am. J. Physiol. Endocrinol. Metab.* **301:** E844–E852.
18. Cameron, N.E. & M.A. Cotter. 2008. Pro-inflammatory mechanisms in diabetic neuropathy: focus on the nuclear factor kappa B pathway. *Curr. Drug Targets* **9:** 60–67.
19. Leininger, G.M., A.M. Vincent & E.L. Feldman. 2004. The role of growth factors in diabetic peripheral neuropathy. *J. Peripher. Nerv. Syst.* **9:** 26–53.
20. Kojima, H., M. Fujimiya, K. Matsumura, *et al.* 2003. NeuroD-betacellulin gene therapy induces islet neogenesis in the liver and reverses diabetes in mice. *Nat. Med.* **9:** 596–603.
21. Kojima, H., M. Fujimiya, K. Matsumura, *et al.* 2004. Extrapancreatic insulin-producing cells in multiple organs in diabetes. *Proc. Natl. Acad. Sci. U.S.A.* **101:** 2458–2463.
22. Fujimiya, M., H. Kojima, M. Ichinose, *et al.* 2007. Fusion of proinsulin-producing bone marrow-derived cells with

hepatocytes in diabetes. *Proc. Natl. Acad. Sci. U.S.A.* **104:** 4030–4035.

23. Oh, S.-H., T.M. Muzzonigro, S.-H. Bae, *et al.* 2004. Adult bone marrow-derived cells trans-differentiating into insulin-producing cells for the treatment of type I diabetes. *Lab. Invest.* **84:** 607–617.

24. Alvarez-Dolado, M., R. Pardal, J.M. Garcia-Verdugo, *et al.* 2003. Fusion of bone-marrow-derived cells with Purkinje neurons, cardiomyocytes and hepatocytes. *Nature* **425:** 968–973.

25. Vassilopoulos, G., P.-R. Wang & D.W. Russell. 2003. Transplanted bone marrow regenerates liver by cell fusion. *Nature* **422:** 901–904.

26. Wang, X., H. Willenbring, Y. Akkari, *et al.* 2003. Cell fusion is the principal source of bone-marrow-derived hepatocytes. *Nature* **422:** 897–900.

27. Nygren, J.M., S. Jovinge, M. Breitbach, *et al.* 2004. Bone marrow-derived hematopoietic cells generate cardiomyocytes at a low frequency through cell fusion, but not transdifferentiation. *Nat. Med.* **10:** 494–501.

28. Rizvi, A.Z., J.R. Swain, P.S. Davies, *et al.* 2006. Bone marrow-derived cells fuse with normal and transformed intestinal stem cells. *Proc. Natl. Acad. Sci. U.S.A.* **103:** 6321–6325.

29. Davies, P.S., A.E. Powell, J.R. Swain & M.H. Wong. 2009. Inflammation and proliferation act together to mediate intestinal cell fusion. *PLoS One* **4:** e6530.

30. Weimann, J.M., C.B. Jahansson, A. Trejo & H.M. Blau. 2003. Stable reprogrammed heterokaryons form spontaneously in Purkinje Neurons after bone marrow transplant. *Nat. Cell. Biol.* **5:** 959–966.

31. Magrassi, L., P. Grimaldi, A. Ibatici, *et al.* 2007. Induction and survival of binucleated Purkinje neurons by selective damage and aging. *J. Neurosci.* **27:** 9885–9892.

32. Johansson, C.B., S. Youssef, K. Koleckar, *et al.* 2008. Extensive fusion of haematopoietic cells with Purkinje neurons in response to chronic inflammation. *Nat. Cell. Biol.* **10:** 575–583.

33. Nygren, J.M., K. Liuba, M. Breitbach, *et al.* 2008. Myeloid and lymphoid contribution to non-haematopoietic lineages through irradiation-induced heterotypic cell fusion. *Nat. Cell. Biol.* **10:** 584–592.

34. Hara, M., X. Wang, T. Kawamura, *et al.* 2003. Transgenic mice with green fluorescent protein-labeled pancreatic b-cells. *Am. J. Physiol. Endocrinol. Metab.* **284:** e177–e183.

35. Terashima, T., H. Kojima, M. Fujimiya, *et al.* 2005. The fusion of bone-marrow-derived proinsulin-expressing cells with nerve cells underlies diabetic neuropathy. *Proc. Natl. Acad. Sci. U.S.A.* **102:** 12525–12530.

36. Terashima T., H. Kojima & L. Chan. 2011. Bone marrow expression of poly(ADP-Ribose) polymerase underlies diabetic neuropathy via hematopoietic-neuronal cell fusion. *FASEB J.* [Epub ahead of print].

37. Brownlee, M. 2005. The pathobiology of diabetic complications—a unifying mechanism. *Diabetes* **54:** 1615–1625.

38. Edwards, J.L., A.M. Vincent, H.T. Cheng & E.L. Feldman. 2008. Diabetic neuropathy: mechanisms to management. *Pharmacol. Ther.* **120:** 1–34.

39. Zochodne, D.W. 2008. Diabetic polyneuropathy: an update. *Curr. Opin. Neurol.* **21:** 527–533.

40. Yasuda, H., M. Terada, K. Maeda, *et al.* 2003. Diabetic neuropathy and nerve regeneration. *Prog. Neurobiol.* **69:** 229–285.

41. Singec, I. & E.Y. Snyder. 2008. Inflammation as a matchmaker: revisiting cell fusion. *Nat. Cell. Biol.* **10:** 503–505.

42. Larsson, L.I., B. Bjerregaard & J.F. Talts. 2008. Cell fusions in mammals. *Histochem. Cell. Biol.* **129:** 551–561.

43. Lluis, F. & M.P. Cosma. 2010. Cell-fusion-mediated somatic-cell reprogramming: a mechanism for tissue regeneration. *J. Cell. Physiol.* **223:** 6–13.

44. Obrosova, I.G., V.R. Drel, P. Pacher, *et al.* 2005. Oxidative-nitrosative stress and poly(ADP-Ribose) polymerase (PARP) activation in experimental diabetic neuropathy: the relation is revisited. *Diabetes* **54:** 3435–3441.

45. Szabo, C. 2005. Roles of poly(ADP-ribose) polymerase activation in the pathogenesis of diabetes mellitus and its complications. *Pharmacol. Res.* **52:** 60–71.

46. Southan, G.J. & C. Szabo. 2003. Poly(ADP-ribose) polymerase inhibitors. *Curr. Med. Chem.* **10:** 321–340.

47. Figueroa-Romero, C., M. Sadidi & E.L. Feldman. 2008. Mechanisms of disease: the oxidative stress theory of diabetic neuropathy. *Rev. Endocr. Metab. Disord.* **9:** 301–314.

48. Li, F., V.R. Drel, C. Szabo, *et al.* 2005. Low-dose poly(ADP-Ribose) polymerase inhibitor-containing combination therapies reverse early peripheral diabetic neuropathy. *Diabetes* **54:** 1514–1522.

49. Ilnytska, O., V.V. Lyzogubov, M.J. Stevens, *et al.* 2006. Poly(ADP-ribose) polymerase inhibition alleviates experimental diabetic sensory neuropathy. *Diabetes* **55:** 1686–1694.

50. Drel, V.R., S. Lupachyk, H. Shevalye, *et al.* 2010. New therapeutic and biomarker discovery for peripheral diabetic neuropathy: PARP inhibitor, nitrotyrosine, and tumor necrosis factor-{alpha}. *Endocrinology* **151:** 2547–2555.

51. Obrosova, I.G., F. Li, O.I. Abatan, *et al.* 2004. Role of poly(ADP-ribose) polymerase activation in diabetic neuropathy. *Diabetes* **53:** 711–720.

52. Obrosova, I.G. 2009. Diabetes and the peripheral nerve. *Biochim. Biophys. Acta* **1792:** 931–940.

53. Clemens T.L. & G. Karsenty. 2011. The osteoblast: an insulin target cell controlling glucose homeostasis. *J. Bone Miner. Res.* **26:** 677–680.

Ann. N.Y. Acad. Sci. ISSN 0077-8923

ANNALS OF THE NEW YORK ACADEMY OF SCIENCES

Issue: *Skeletal Biology and Medicine II*

Noninvasive imaging of bone microarchitecture

Janina M. Patsch,[1,2] Andrew J. Burghardt,[1] Galateia Kazakia,[1] and Sharmila Majumdar[1]

[1]Musculoskeletal Quantitative Imaging Research, Department of Radiology and Biomedical Imaging, University of California, San Francisco, California. [2]Division of Neuroradiology and Musculoskeletal Radiology, Department of Radiology, Medical University of Vienna, Vienna, Austria

Address for correspondence: Janina Patsch, M.D., 185 Berry Street, Suite 350, San Francisco, CA, 94107. janina.patsch@ucsf.edu

The noninvasive quantification of peripheral compartment-specific bone microarchitecture is feasible with high-resolution peripheral quantitative computed tomography (HR-pQCT) and high-resolution magnetic resonance imaging (HR-MRI). In addition to classic morphometric indices, both techniques provide a suitable basis for virtual biomechanical testing using finite element (FE) analyses. Methodical limitations, morphometric parameter definition, and motion artifacts have to be considered to achieve optimal data interpretation from imaging studies. With increasing availability of *in vivo* high-resolution bone imaging techniques, special emphasis should be put on quality control including multicenter, cross-site validations. Importantly, conclusions from interventional studies investigating the effects of antiosteoporotic drugs on bone microarchitecture should be drawn with care, ideally involving imaging scientists, translational researchers, and clinicians.

Keywords: high-resolution peripheral quantitative computed tomography (HR-pQCT); high-resolution magnetic resonance imaging (HR-MRI); bone microarchitecture; bone quality

Introduction

The term *bone strength* describes the ability of the skeleton to resist fractures. For the last several decades, bone mineral density (BMD) has been used as the main surrogate for bone strength. The definitions of osteoporosis and clinical decision making have been strongly influenced by this density-based approach, and dual X-ray absorptiometry (DXA) is clearly the current clinical gold standard technique for bone density assessment. DXA is widely available, provides good precision and reproducibility, but does not offer insight into compartment-specific bone densities. Being a projectional technique, DXA underestimates the density of smaller bones that play a role when assessing BMD in children and with regards to gender-specific studies.[1] In addition, bone strength is independently modulated by a wide spectrum of tissue-based measures, including bone microarchitecture, bone turnover, bone matrix status (e.g., collagen properties and mineralization), as well as accumulated microdamage.[2] All these factors have been often comprehensively termed *bone quality*. In the past, most of these features were determined using histology and bone biopsies, and while that is still the case for some of the parameters, during the past 20 years, novel high-resolution imaging techniques that are capable of noninvasive assessment of bone microarchitecture have emerged. These techniques include high-resolution peripheral quantitative computed tomography (HR-pQCT) and high-resolution magnetic resonance imaging of bone (HR-MRI). In the following sections a few of the methods for the noninvasive assessment of bone microarchitecture will be reviewed.

High-resolution peripheral quantitative computed tomography

High-resolution peripheral quantitative computed tomography (HR-pQCT) imaging requires a dedicated extremity scanner. Using HR-pQCT, measures of three-dimensional (3D) bone geometry, overall and compartment-specific bone density and bone microarchitecture can be acquired within a scan time of 3 minutes. To obtain a 3D stack of images with an isotropic voxel size of 82 μm (110 slices,

doi: 10.1111/j.1749-6632.2011.06282.x

9 mm scan length), the patient radiation exposure is ~4 μSv, which is about 500–1000 fold smaller compared to a clinical abdominal CT scan.[3] The standard imaging region is located proximal to the radiocarpal joint at the ultradistal radius and proximal to the ankle joint at the ultradistal tibia. While limits of the scanner geometry constrain the range that is possible to scan, acquisition protocols can be modified to specifically capture more proximal zones dominated by cortical bone or more distal zones to capture joints/metacarpals/metatarsals. Nevertheless, imaging mid-diaphyseal regions of the adult distal radius and tibia or more proximal long bones sites are not possible at present.

HR-pQCT image analysis

For the evaluation of quantitative measures, the periosteal contours are drawn in a semiautomatic manner. Prior to the application to a fixed segmentation threshold, a smoothing and edge enhancement procedure is performed on grayscale images.[4] HR-pQCT does not only provide a global, three-dimensional, volumetric BMD but also separately yields compartment-specific bone densities for the trabecular core and the cortical shell. Global, volumetric BMD is convertible into areal BMD values.[5] BV/TV is derived from trabecular volumetric BMD assuming a fixed mineralization of 1200 mg HA/cm^3 for compact bone. Standard morphometric parameters are analogous to those used in classical static histomorphometry.[6] Trabecular number (Tb.N) and network heterogeneity are directly measured by 3D distance transformations.[7] Trabecular thickness (Tb.Th in micrometers) and trabecular separation (Tb.Sp in micrometers) are derived from Tb.N and/or trabecular density using stereological standard relations assuming a plate-model geometry.[8] Measurement reproducibility has been shown to be higher for density parameters (<1% root mean square coefficient of variance–RMSCV) than for microarchitectural metrics (<4.5% RMSCV).[9]

Validation experiments comparing *ex vivo* HR-pQCT with DXA and micro-computed tomography (μCT) at resolutions ranging from 10 to 25 μm (the gold standard) yielded strong correlations between the methods, but also stressed resolution dependence.[9–12]

HR-pQCT also provides measures of cortical bone properties.[13] The noninvasive assessment of cortical bone density and quality is highly clinically relevant as cortical integrity has been shown to be a crucial determinant of overall bone strength.[14,15] Especially cortical porosity is increasingly recognized as a pathomorphologic surrogate of poor bone quality.[14,16] However, it remains to be stressed that due to the *in vivo* voxel size of HR-pQCT data, noninvasive imaging captures only relatively large pores (>82 μm).[14]

In addition to cortical bone analyses, region-specific evaluation can enhance the sensitivity of HR-pQCT with regard to age- and gender-related changes.[17] Regional analysis of cortical and trabecular bone subdivided in standardized quadrants was also shown to improve the detectability of longitudinal changes in bone microarchitecture caused by alendronate treatment when compared with global evaluations.[18]

Various other postprocessing techniques including finite element (FE) modeling are applicable to HR-pQCT data. FE modeling is a biomechanical computation method that yields loading scenario-specific, image-based estimates of bone strength (e.g., ultimate force required to fracture upon a certain type of fall). Prior to the introduction of HR-pQCT, FE analyses have been performed with central QCT (hip and lumbar spine) and μCT data.[19–21] QCT-derived FE models have been used to monitor antiosteoporotic drug effects in clinical trials.[22–26] HR-pQCT-based FE analyses can been carried out on the entire scan volume as well as a defined subvolume.[27–29] Both have been shown to discriminate men and postmenopausal women with and without fragility fractures independent of BMD.[30,31]

Another specific image evaluation algorithm, termed *individual trabecula segmentation* (ITS), provides insight into trabecular topological characteristics by separately quantifying plate- and rod-like structures.[32] This technique also describes the axial bone volume fraction, that is, the number of trabecular structures that are aligned with the longitudinal bone/scan axis. ITS-based metrics were shown to display high correlations with FE-derived biomechanical indices.[33]

Motion artifacts play a significant role in impairing image quality and parameter calculations. They are more frequent and more severe in radius than tibia scans. Motion affects morphometric indices to a larger extent than density measures. Currently, motion artifacts are subjectively evaluated on the basis of a semiquantitative, visual grading scale

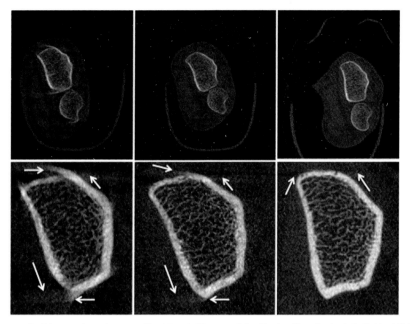

Figure 1. Influence of subject motion of scan-quality HR-pQCT scans of the distal radius of a single individual (left: grade 5—not usable for analysis; middle: grade 3—still usable; right: grade 2—almost motionless, good image quality). Arrows point at artifact streaks induced by rotational motion.

(Fig. 1).[34] While there has been some effort to develop fully quantitative estimates of patient motion, more optimization and validation is warranted to translate these tools to routine use in clinical research studies.[35,36] Of practical relevance, an optimized patient set-up with detailed explanations highlighting the importance of avoiding motion during the scan, proper arm and leg fixation and a quiet room atmosphere, will help to reduce subject motion.

Besides resolution and motion, beam-hardening and X-ray scatter effects can also impair HR-pQCT-derived measures. Accordingly, variations in trabecular bone density and cortical thickness can introduce a relevant bias in cross-sectional studies.[37]

Comparison with iliac crest biopsies, DXA, and QCT

For decades, histology and histomorphometry of iliac crest biopsies have represented the gold-standard techniques for the assessment of bone microarchitecture. Compared with bone biopsies, noninvasiveness and the possibility of multiple local follow-ups are major advantages of HR-pQCT. However, HR-pQCT can only quantify mineralized bone properties but not osteoid or bone cell characteristics. Conclusions regarding global or focal mineralization

defects such as osteomalacia or secondary causes of osteoporosis including malignant bone marrow infiltration (e.g., by mast cells or lymphoma cells) cannot be drawn without invasive sampling. Thus clinical biopsy indications remain unchanged for certain applications or for investigating the aspects of bone health not captured by imaging. Comparing iliac crest bone biopsies with *in vivo* HR-pQCT of the radius, Cohen *et al.* found significant but modest associations between measures of microarchitecture and stiffness obtained from both methods.[38]

Neither iliac crest biopsies nor HR-pQCT imaging provide direct information on bone quality of the central skeleton. Since the ability to draw reliable conclusions from peripheral bone microarchitecture to central bone quality is of major clinical interest, several studies have investigated these relations. For the radius, Sornay-Rendu reported good agreement between local aBMD by DXA and densities and microstructure obtained by HR-pQCT. Correlations were also significant but slightly weaker for HR-pQCT of the distal tibia and DXA of the hip.[39] Vico *et al.* also found significant associations of HR-pQCT of the radius and tibia with spine and hip BMD by DXA. Stressing the fact that the spine mainly represents a trabecular bone site, they reported correlations with cortical parameters to be weak at the radius and event absent at the tibia.[40] More recently, Liu *et al.* demonstrated that HR-pQCT-based measurements of bone density, bone microstructure and FE analyses at both sites are significantly related to QCT/FE-derived biomechanical properties of the lumbar spine and hip.[41]

Cross-sectional studies

Highlighting that noninvasive assessment of bone microarchitecture could add to the refinement of clinical fracture risk prediction, HR-pQCT has been shown to differentiate patients with and without prevalent fragility fractures irrespective of their BMD. Boutroy *et al.* found significantly lower trabecular density (−12.3%), lower trabecular number (−8.5%) and higher standard deviations in trabecular separation (+25.6%) in osteopenic women with fractures when compared to nonfracture subjects with identical BMD. Fracture history did, but DXA alone did not adequately depict impaired bone health.[42] Poor peripheral bone microarchitecture was also a characteristic feature in male and female patients with spine, hip and peripheral fractures.[40,43,44] One study showed that with increasing grade of spine fracture severity, HR-pQCT parameters also deteriorated stepwise.[39]

Cross-sectional HR-pQCT studies have also provided insight into age- and gender-specific aspects of bone quality.[16,45,46] Compartment-specifc analyses of bone density and microarchitecture seem of particular interest in clarifying osteologic paradoxes with discrepancies between fracture prevalence and mere densitometric risk prediction (e.g., lower fracture rates in Asians or elevated fracture risk in diabetic bone disease): Asian men and women have smaller bones; thus areal BMD as measured by DXA tends to underestimate their real bone density. Nevertheless, Asians sustain fewer fractures. Using HR-pQCT, Wang *et al.* found that in spite of relatively low total bone area, premenopausal Asian women displayed significantly thicker cortices and a richer trabecular microarchitecture than Caucasians.[47] Accordingly, FE analyses yielded higher estimates of bone stiffness/strength.[48] Menopause diminishes some of these microstructural advantages but significant racial differences remain detectable.[49] However, differences in bone length are currently an unaccounted bias in HR-pQCT studies that investigate gender- or race-specific bone char-

acteristics. Since the standard region of interest is currently positioned with a fixed offset with regard to anatomical landmarks of the wrist and ankle joint irrespective of total bone length, radii, and tibiae of taller subjects (e.g., men versus women) are scanned relatively more distal than shorter bones. This could confound differences found for densities and morphometric parameters.

Patients with type 2 diabetes mellitus present with an opposite clinical situation. DXA-based risk assessment typically attributes normal to high BMD to diabetics, nevertheless they were shown to fracture more often than age- and gender-matched nondiabetics.[50] Burghardt *et al.* have suggested that this increased fracture risk in individuals with type 2 diabetes mellitus may be attributable to ultrastructural deficits in cortical bone.[51] However, in accordance with older studies using DXA, they also reported normal to slightly elevated trabcular BMD in diabetic subjects. Somehow contrasting these data, another recent HR-pQCT study found almost similar bone microarchitecture in postmenopausal women with and with diabetes but Shu *et al.* did not include fracture patients and cortical bone has not been specifically analyzed for porosity.[52]

Moreover, HR-pQCT has been used to address gender-specific research questions.[46] Although it had been repeatedly postulated that osteoporotic women displayed different alterations in bone microarchitecture than men, the actual evidence for a specific male microstructural disease pattern remained limited for a long time.[53,54] Using an HR-pQCT prototype, Khosla *et al.* were the first to noninvasively address gender differences in bone microstructure.[45] In accordance with previous biopsy studies they found that BV/TV and Tb.Th. were significantly higher in young men than in young women. They further stated that the rate of age-related decline in trabecular bone volume seemed independent of gender, but also reaffirmed that in aging but otherwise healthy men trabecular thinning seemed to predominate over actual loss of trabeculae. Larger bones, greater periosteal apposition upon aging, conservation of trabecular structures and the lack of a real menopause equivalent contribute to lower fracture risk in men.[55]

Bone properties of patients with renal osteodystrophy and primary hyperparathyroidism have also been characterized by HR-pQCT.[56–59] The use of HR-pQCT is specifically promising in patients with

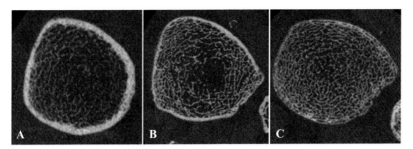

Figure 2. Three examples of HR-pQCT scans of the distal tibia. (A) Normal premenopausal control. (B) Postmenopausal women with chronic kidney disease. Morphologic features include cortical thinning, loss of trabeculae, and severe arterial calcifications. (C) Postmenopausal and age-related cortical porosity with relative preservation of the trabecular compartment.

chronic kidney disease (CKD; Fig. 2). These patients are particularly prone to fractures but DXA often remains false negative due to vascular calcifications that are erroneously integrated in projectional bone density. In addition, a compartment-specific approach appears favorable in CKD patients because secondary or tertiary hyperparathyroidism is predominantly leading to cortical bone loss.

Longitudinal and multicenter HR-pQCT studies

As with DXA before it, the utility of HR-pQCT to address important questions related to the epidemiology of osteoporosis, therapeutic antifracture efficacy, and the development of normative databases for clinical assessment of skeletal health requires scalability to standardized multicenter data pools. To fully realize the potential of HR-pQCT to investigate the role of bone quality in these contexts, appropriate tools are needed to characterize long-term and interscanner sources of variability, evaluate new acquisition and postprocessing techniques designed to improve comparability longitudinally and amongst scanners, and develop appropriate cross-calibration procedures to standardize the quantitative outcomes. Burghardt *et al.* proposed the use of structure- and composition-realistic anthropomorphic phantoms constructed from static cadaveric bone tissue to monitor longitudinal stability of HR-pQCT measures and characterize interscanner variability.[60] Based on phantom measurements at six imaging centers, they reported interscanner variability (RMSCV%) comparable in magnitude to short term *in vivo* reproducibility reported elsewhere.[9,61] Densitometric measures had the highest interscanner reproducibility (approximately 1%)

while geometric and microstructural measures were moderately less precise (4–6%). The sources of error were found to be variable and specific to each individual scanner, including differences in noise or resolution performance, the geometric and density calibration, or related to postprocessing factors.

Among other caveats in longitudinal studies, the assumption of a fixed matrix mineralization for HR-pQCT-based bone assessment has to be considered. Many drugs with proven antifracture efficacy significantly affect tissue mineralization.[62,63] Irrespective of drug effects on bone volume, such changes in mineralization would be expected to cause an additional increase in volumetric BMD (vBMD). Due to derivation of BV/TV, Tb.Th and Tb.Sp from trabecular vBMD, this could confound interpretation of temporal trends.

Somewhat surprisingly, longitudinal HR-pQCT data interpretation has proven to be challenging.[18,64–67] Burghardt *et al.* followed alendronate versus placebo-treated osteopenic postmenopausal women for two years and found treatment-related increases in tibial cortical area and thickness without changes in cortical density. At the same time, global trabecular bone loss was prevented. As opposed to the tibia as a weight-bearing bone, no positive treatment-related changes were detectable in the radius.[18] Few studies have used data acquired at multiple imaging centers: Seeman *et al.* reported the results of a multicenter, 12-month placebo-controlled trial of denosumab and alendronate in postmenopausal women.[67] In this study, the failure to detect significant longitudinal changes in microstructural measures, despite pronounced densitometric treatment effects, may have been partially due to unaccounted intersite variability.

High-resolution magnetic resonance imaging

As opposed to computed tomography, magnetic resonance imaging (MRI) lacks ionizing radiation. Besides providing significantly better insight to soft tissue properties than CT, the mechanisms of image generation differ significantly between these two modalities. While mineralized tissues attenuate X-rays and thereby finally generate direct image contrast in CT, MRI of bone microstructure is based on a negative image effect (i.e., the missing signal from bone versus the intense signal arising from the marrow). Differing electromagnetic properties—that is, different magnetic susceptibility—are the physical background of tissue-specific signal generation. The mix of trabecular bone and bone marrow causes magnetic field inhomogeneities that are detectable by MRI. These inhomogeneities depend on the number of bone-bone marrow interfaces, the size of individual trabeculae but also on the strength of the static magnetic field.[68–70] As a consequence nuclear spins are dephased and the MR signal is decayed (which is referred to as T2*).[71] T2* decay characteristics of spine but also hip bone marrow were shown to differentiate postmenopausal women with manifest osteoporosis from nonfracture controls.[71,72] From these first clinically relevant steps, bone MRI has developed into an accurate and reproducible high-resolution method with reasonable acquisition time providing detailed insight to bone microstructure. Like HR-pQCT that is technically limited to the distal extremities, HR-MRI has mostly been performed in the peripheral skeleton. However, using SNR efficient sequences, high magnetic field strength (3 Tesla), and phased array coils, high-resolution magnetic resonance imaging has been shown to be feasible at the proximal femur.[73,74] In general, higher field strengths contribute significantly to increased performance of HR-MRI. In an *ex vivo* validation study, Phan *et al.* demonstrated higher correlations for trabecular parameters at 3T than at 1.5T when compared with μ-CT as the gold standard.[74]

HR-MRI of trabecular bone

Based on signal differences of trabecular bone (low signal, i.e., dark striations) and bone marrow (high intertrabecular signal, i.e., bright zones), structural indices, such as app. Tb.N, app. Tb.Th, and app. Tb.Sp, and distance transformation techniques, can be applied to three-dimensional MR images.[75] Further, the ratio of surface voxels (representing trabecular plates) to curve voxels (representing trabecular rods) and the erosion index can be computed. Indicating the predominance of more plate-like trabeculae, a higher surface to curve-ratio is found with more intact bone microstructure. Figure 3 shows an example of a high-resolution MR image obtained at 3 Tesla. Several *ex vivo* validation studies relating MR-derived measures of trabecular structure with bone strength have been conducted.[76–79] The relationship between whole bone strength and bone structure measures has been demonstrated in radii and in the proximal femur.[80,81]

A number of cross-sectional and longitudinal HR-MRI studies have been conducted. In line with lower spinal and trabecular radial BMD, MR-derived trabecular bone volume fraction, trabecular spacing and trabecular number showed significant differences in postmenopausal osteoporotic women with and without fractures when compared with premenopausal healthy controls.[82] Other studies specifically enrolled women with hip fractures and found significantly disturbed trabecular microstructure including low trabecular number at the radius[83] but also the calcaneus.[84] Hypogonadal men with secondary osteoporosis were also studied by HR-MRI and displayed a significantly higher erosion index (+36%) and a significantly lower surface-to-curve ratio (−36%) than eugonadal controls.[85]

Examples for the longitudinal use of HR-MRI include research studies published by Benito *et al.*[86] and Chestnut *et al.*[87] Ninety-one postmenopausal osteoporotic women (n = 46 for nasal spray calcitonin, n = 45 for placebo) were followed for two years. Repeated MRI assessment of trabecular microarchitecture at individual regions of the distal radius revealed preservation (i.e., no significant loss) of structures in the treated group compared with significant deterioration in the placebo control group.[87] Benito *et al.* reported positive effects of testosterone replacement on bone microstructure in ten men with severe hypogonadism after 24 months of treatment.[86]

MRI of cortical bone

MR can also be used to image cortical bone, specifically the proximal femur. The ability of MRI to align the image plane perpendicular to the femoral neck

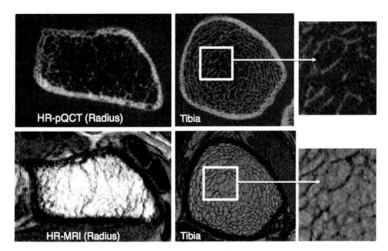

Figure 3. Comparison of HR-pQCT and HR-MRI in an osteopenic, early postmenopausal woman. Upper row: HR-pQCT of the distal radius and tibia. HR-MRI at identical skeletal sites (below) depicts trabecular structures by inverse contrast.

is a great advantage as it enables image acquisition with accurate cortical boundary definitions.[88] Besides cortical morphometry (e.g., quantification of cortical thickness), MRI allows the visualization of soft tissues. Complementing HR-pQCT-based morphometric and biomechanical approaches to cortical porosity, HR-MRI has been used to characterize the tissue properties of pore contents (e.g., marrow fat versus vasculature).[89] Using advanced MRI methods with ultra short echo times (UTE), cortical bone water content in the microscopic pores of the Haversian canals and the lacuno-canalicular systems can be quantified in addition.[90,91] This is of relevance because a significant number of cortical pores is too small a number to be directly detected even by high-resolution imaging techniques. Thus quantification of cortical bone water could provide a potential indirect surrogate measure of bone porosity without resolving these individual small pores. The *in vivo* quantification of cortical bone water revealed a 65% increase in postmenopausal versus premenopausal women.[92] Patients with renal osteodystrophy had even more pronounced increases (+135% compared with controls). In this study, cortical bone water was shown to be a highly sensitive parameter with group differences exceeding those of DXA.

Additional applications of MRI in bone research

Besides high-resolution imaging, MRI can complement bone research by its capability to quantify bone and bone marrow properties on a biochemical level. Using bone marrow spectroscopy, the content of fat, water, and other metabolites can be assessed quantitatively within a defined volume of interest.[93] In the light of the emerging importance of the interactions between marrow fat and bone, the possibility of noninvasive characterization of bone marrow composition is being increasingly recognized as an independent surrogate for bone quality.[94–96] Besides evidence for the negative effects of increases in marrow fat content, fat composition appears to be linked with bone density.[97]

Summary

In conclusion, HR-pQCT and HR-MRI provide novel, noninvasive and accurate options for the quantification of bone microarchitecture. Although current research focusing on potential clinical applications seems promising, both techniques are still limited to research use. Conclusions from high-resolution bone imaging studies should only be drawn with recognition of technical caveats. Moreover, quality control should be standardized.

Acknowledgement

Erwin-Schrödinger Grant No. J-3079 (Austrian Science Fund) to J.M.P.

Conflicts of interest

The authors declare no conflict of interest.

References

1. Lewiecki, E.M., S. Baim & N. Binkley. 2008. Report of the International Society for Clinical Densitometry 2007 Adult Position Development Conference and Official Positions. *South Med. J.* **101:** 735–739.

2. Anonymous. 2001. NIH Consensus Development Panel on Osteoporosis Prevention and Therapy. *JAMA* **285:** 785–795.

3. Damilakis, J., J.E. Adams, G. Guglielmi, *et al.* 2010. Radiation exposure in X-ray-based imaging techniques used in osteoporosis. *Eur. Radiol.* **20:** 2707–2714.

4. Laib, A., P. Ruegsegger. 1999. Comparison of structure extraction methods for *in vivo* trabecular bone measurements. *Comput. Med. Imaging Graph.* **23:** 69–74.

5. Burghardt, A.J., G.J. Kazakia T.M. Link, *et al.* 2009. Automated simulation of areal bone mineral density assessment in the distal radius from high-resolution peripheral quantitative computed tomography. *Osteoporos. Int.* **20:** 2017–2024.

6. Parfitt, A.M., M.K. Drezner, F.H. Glorieux, *et al.* 1987. Bone histomorphometry: standardization of nomenclature, symbols, and units. Report of the ASBMR Histomorphometry Nomenclature Committee. *J. Bone Miner. Res.* **2:** 595–610.

7. Hildebrand, T. & P. Ruegsegger. 1997. A new method for the model-independent assessment of thickness in three-dimensional images. *J. Microsc.* **185:** 67–75.

8. Laib, A., H.J. Hauselmann & HP. Ruegsegger. 1998. *In vivo* high resolution 3D-QCT of the human forearm. *Technol. Health Care* **6:** 329–337.

9. MacNeil, J.A. & S.K. Boyd. 2007. Accuracy of high-resolution peripheral quantitative computed tomography for measurement of bone quality. *Med. Eng. Phys.* **29:** 1096–1105.

10. Macneil, J.A. & S.K. Boyd. 2008. Bone strength at the distal radius can be estimated from high-resolution peripheral quantitative computed tomography and the finite element method. *Bone* **42:** 1203–1213.

11. Sode, M., A.J. Burghardt, R.A. Nissenson, *et al.* 2008. Resolution dependence of the non-metric trabecular structure indices. *Bone* **42:** 728–736.

12. Liu, X.S., X.H. Zhang, K.K. Sekhon, *et al.* 2010. High-resolution peripheral quantitative computed tomography can assess microstructural and mechanical properties of human distal tibial bone. *J. Bone Miner. Res.* **25:** 746–756.

13. Burghardt, A.J., H.R. Buie, A. Laib, *et al.* 2010. Reproducibility of direct quantitative measures of cortical bone microarchitecture of the distal radius and tibia by HR-pQCT. *Bone* **47:** 519–528.

14. Nishiyama, K.K., H.M. Macdonald, H.R. Buie, *et al.* 2010. Postmenopausal women with osteopenia have higher cortical porosity and thinner cortices at the distal radius and tibia than women with normal aBMD: an *in vivo* HR-pQCT study. *J. Bone Miner. Res.* **25:** 882–890.

15. Holzer, G., G. von Skrbensky, L.A. Holzer, *et al.* 2009. Hip fractures and the contribution of cortical versus trabecular bone to femoral neck strength. *J. Bone Miner. Res.* **24:** 468–474.

16. Burghardt, A.J., G.J. Kazakia, S. Ramachandran, *et al.* 2010. Age- and gender-related differences in the geometric properties and biomechanical significance of intracortical porosity in the distal radius and tibia. *J. Bone Miner. Res.* **25:** 983–993.

17. Sode, M., A.J. Burghardt, G. Kazakia, *et al.* 2010. Regional variations of gender-specific and age-related differences in trabecular bone structure of the distal radius and tibia. *Bone* **46:** 1652–1660.

18. Burghardt, A.J., G. Kazakia, M. Sode, *et al.* 2010. A longitudinal HR-pQCT study of alendronate treatment in postmenopausal women with low bone density: relations among density, cortical and trabecular microarchitecture, biomechanics, and bone turnover. *J. Bone Miner. Res.* **25:** 2558–2571.

19. Keaveny, T.M. 2010. Biomechanical computed tomography-noninvasive bone strength analysis using clinical computed tomography scans. *Ann. N. Y. Acad. Sci.* **1192:** 57–65.

20. van Rietbergen, B., H. Weinans, R. Huiskes, *et al.* 1995. A new method to determine trabecular bone elastic properties and loading using micromechanical finite-element models. *J. Biomech.* **28:** 69–81.

21. van Rietbergen, B. 2001. Micro-FE analyses of bone: state of the art. *Adv. Exp. Med. Biol.* **496:** 21–30.

22. Black, D.M., S.L. Greenspan, K.E. Ensrud, *et al.* 2003. The effects of parathyroid hormone and alendronate alone or in combination in postmenopausal osteoporosis. *N. Engl. J. Med.* **349:** 1207–1215.

23. Black, D.M., J.P. Bilezikian, K.E. Ensrud, *et al.* 2005. One year of alendronate after one year of parathyroid hormone (1–84) for osteoporosis. *N. Engl. J. Med.* **353:** 555–565.

24. Graeff, C., Y. Chevalier, M. Charlebois, *et al.* 2009. Improvements in vertebral body strength under teriparatide treatment assessed *in vivo* by finite element analysis: results from the EUROFORS study. *J. Bone Miner. Res.* 2009 **24:** 1672–1680.

25. Eastell, R., T. Lang, S. Boonen, *et al.* 2010. Effect of once-yearly zoledronic acid on the spine and hip as measured by quantitative computed tomography: results of the HORIZON Pivotal Fracture Trial. *Osteoporos. Int.* **21:** 1277–1285.

26. Genant, H.K., K. Engelke, D.A. Hanley, *et al.* 2010. Denosumab improves density and strength parameters as measured by QCT of the radius in postmenopausal women with low bone mineral density. *Bone* **47:** 131–139.

27. CAVE: 2xLiu, X.S., X.H. Zhang, K.K. Sekhon, *et al.* 2010. High-resolution peripheral quantitative computed tomography can assess microstructural and mechanical properties of human distal tibial bone. *J. Bone Miner. Res.* **25:** 746–756.

28. Macneil, J.A. & S.K. Boyd. 2008. Bone strength at the distal radius can be estimated from high-resolution peripheral quantitative computed tomography and the finite element method. *Bone* **42:** 1203–1213.

29. Mueller, T.L., D. Christen, S. Sandercott, *et al.* 2011. Computational finite element bone mechanics accurately predicts mechanical competence in the human radius of an elderly population. *Bone* **4:** 1232–1238.

30. Boutroy, S., B. Van Rietbergen, E. Sornay-Rendu, *et al.* 2008. Finite element analysis based on in vivo HR-pQCT images of the distal radius is associated with wrist fracture in postmenopausal women. *J. Bone Miner. Res.* **23:** 392–399.

31. Vilayphiou, N., S. Boutroy, P. Szulc, *et al.* 2011. Finite element analysis performed on radius and tibia HR-pQCT images and fragility fractures at all sites in men. *J. Bone Miner. Res.* **26:** 965–973.

32. Liu, X.S., A. Cohen, E. Shane, *et al*. 2010. Individual trabeculae segmentation (ITS)-based morphological analyses of high resolution peripheral quantitative computed tomography images detect abnormal trabecular plate and rod microarchitecture in premenopausal women with idiopathic osteoporosis. *J. Bone Miner. Res.* **25**: 1486–1505.

33. Liu, X.S., E. Shane, D.J. McMahon, *et al*. 2011. Individual trabeculae segmentation (ITS)-based morphological analysis of micro-scale images of human tibial trabecular bone at limited spatial resolution. *J. Bone Miner. Res.* **26**: 2184–2193.

34. Pialat, J.B., A.J. Burghardt, M. Sode, *et al*. 2010. Motion artifacts in high-resolution peripheral quantitative computed tomography of wrist and ankle: usefulness of visual grading to assess image quality. *J. Bone Miner. Res.* **25**: S373.

35. Sode, M., A.J. Burghardt, J.B. Pialat, *et al*. 2011. Quantitative characterization of subject motion in HR-pQCT images of the distal radius and tibia. *Bone* **48**: 1291–1297.

36. Pauchard, Y., F.J. Ayres & Boyd S.K. 2011. Automated quantification of three-dimensional subject motion to monitor image quality in high-resolution peripheral quantitative computed tomography. *Phys. Med. Biol.* **56**: 6523–6543. [Epub ahead of print]

37. Sekhon, K., G. Kazakia, A.J. Burghardt, *et al*. 2009. Accuracy of volumetric bone mineral density measurement in high-resolution peripheral quantitative computed tomography. *Bone* **45**: 473–479.

38. Cohen, A., D.W. Dempster, R. Müller, *et al*. 2010. Assessment of trabecular and cortical architecture and mechanical competence of bone by high-resolution peripheral computed tomography: comparison with transiliac bone biopsy. *Osteoporos. Int.* **21**: 263–273.

39. Sornay-Rendu, E., S. Boutroy, F. Munoz, *et al*. 2007. Alterations of cortical and trabecular architecture are associated with fractures in postmenopausal women, partially independent of decreased BMD measured by DXA: the OFELY study. *J. Bone Miner. Res.* **22**: 425–433.

40. Vico, L., M. Zouch, A. Amirouche, *et al*. 2008. High-resolution pQCT analysis at the distal radius and tibia discriminates patients with recent wrist and femoral neck fractures. *J. Bone Miner. Res.* **23**: 1741–1750.

41. Liu, X.S., A. Cohen, E. Shane, *et al*. 2010. Bone density, geometry, microstructure and stiffness: relationships between peripheral and central skeletal sites assessed by DXA, HR-pQCT, and cQCT in premenopausal women. *J. Bone Miner. Res.* **25**: 2229–2238.

42. Boutroy, S., M.L. Bouxsein, F. Munoz, *et al*. 2005. *In vivo* assessment of trabecular bone microarchitecture by high-resolution peripheral quantitative computed tomography. *J. Clin. Endocrinol. Metab.* **90**: 6508–6515.

43. Vilayphiou, N., S. Boutroy, P. Szulc, *et al*. 2011. Finite element analysis performed on radius and tibia HR-pQCT images and fragility fractures at all sites in men. *J. Bone Miner. Res.* **26**: 965–973.

44. Stein, E.M., X.S. Liu, T.L. Nickolas, *et al*. 2011. Abnormal microarchitecture and stiffness in postmenopausal women with ankle fractures. *J. Clin. Endocrinol. Metab.* **96**: 2041–2048.

45. Khosla, S., B.L. Riggs, E.J. Atkinson, *et al*. 2006. Effects of sex and age on bone microstructure at the ultradistal radius: a population-based noninvasive *in vivo* assessment. *J. Bone Miner. Res.* **21**: 124–131.

46. Macdonald, H.M., K.K. Nishiyama, J. Kang, *et al*. 2011. Age-related patterns of trabecular and cortical bone loss differ between sexes and skeletal sites: a population-based HR-pQCT study. *J. Bone Miner. Res.* **26**: 50–62.

47. Wang, X.F., Q. Wang, A. Ghasem-Zadeh, *et al*. 2009. Differences in macro- and microarchitecture of the appendicular skeleton in young Chinese and white women. *J. Bone Miner. Res.* **24**: 1946–1952.

48. Liu, X.S., M.D. Walker, D.J. McMahon, *et al*. 2011. Better skeletal microstructure confers greater mechanical advantages in Chinese–American women versus white women. *J. Bone Miner. Res.* **26**: 1783–1792.

49. Walker, M.D., X.S. Liu, E. Stein, *et al*. 2011. Differences in bone microarchitecture between postmenopausal Chinese-American and white women. *J. Bone Miner. Res.* **26**: 1392–1398.

50. Schwartz, A.V. & D.E. Sellmeyer. 2004. Women, type 2 diabetes, and fracture risk. *Curr. Diab Rep.* **4**: 364–369.

51. Burghardt, A. J., A.S. Issever, A.V. Schwartz, *et al*. 2010. High-resolution peripheral quantitative computed tomographic imaging of cortical and trabecular bone microarchitecture in patients with type 2 diabetes mellitus. *J. Clin. Endocrinol. Metab.* **95**: 5045–5055.

52. Shu, A., M.T. Yin, E. Stein, *et al*. 2011. Bone structure and turnover in type 2 diabetes mellitus. *Osteoporos. Int.* Mar 19 [Epub ahead of print].

53. Aaron, J.E., N.B. Makins & K. Sagreiy. 1987. The microanatomy of trabecular bone loss in normal ageing men and women. *Clin. Orthop.* **215**: 260–271.

54. Seeman, E. 2001. Unresolved issues in osteopororsis in men. *Rev. Endocrinol. Metab. Disord.* **2**: 45–64.

55. Patsch, J.M., J. Deutschmann, P. Pietschmann. 2011. Gender aspects of osteoporosis and bone strength. *Wien. Med. Wochenschr.* **161**: 117–123.

56. Bacchetta, J., S. Boutroy, N. Vilayphiou, *et al*. 2010. Early impairment of trabecular microarchitecture assessed with HR-pQCT in patients with stage II-IV chronic kidney disease. *J. Bone Miner. Res.* **25**: 849–857.

57. Nickolas, T.L., E. Stein, A. Cohen, *et al*. 2010. Bone mass and microarchitecture in CKD patients with fracture. *J. Am. Soc. Nephrol.* **21**: 1371–1380.

58. Cejka, D., A. Jäger-Lansky, H. Kieweg, *et al*. 2011. Sclerostin serum levels correlate positively with bone mineral density and microarchitecture in haemodialysis patients. *Nephrol. Dial. Transplant*. May 25 [Epub ahead of print].

59. Hansen, S., J.E. Beck Jensen, L. Rasmussen, *et al*. 2010. Effects on bone geometry, density, and microarchitecture in the distal radius but not the tibia in women with primary hyperparathyroidism: A case-control study using HR-pQCT. *J. Bone Miner. Res.* **25**: 1941–1947.

60. Burghardt, A.J., B. Hermannsson, J.B. Pialat, *et al*. 2010. Cross-site reproducibility of cortical and trabecular bone density and micro-architecture measurements by HR-pQCT. *Osteoporos. Int.* **21**: S45–S46.

61. Kazakia, G.J., B. Hyun, A.J. Burghardt, *et al*. 2008. In vivo determination of bone structure in postmenopausal women:

a comparison of HR-pQCT and high-field MR imaging. *J. Bone Miner. Res.* **23:** 463–474.

62. Boivin, G.Y., P.M. Chavassieux, A.C. Santora, *et al.* 2000. Alendronate increases bone strength by increasing the mean degree of mineralization of bone tissue in osteoporotic women. *Bone* **27:** 687–694.

63. Roschger, P., I. Manjubala, N. Zoeger, *et al.* 2010. Bone material quality in transiliac bone biopsies of postmenopausal osteoporotic women after 3 years of strontium ranelate treatment. *J. Bone Miner. Res.* **25:** 891–900.

64. Macdonald, H.M., K.K. Nishiyama, D.A. Hanley, *et al.* 2011. Changes in trabecular and cortical bone microarchitecture at peripheral sites associated with 18 months of teriparatide therapy in postmenopausal women with osteoporosis. *Osteoporos. Int.* **22:** 357–362.

65. Rizzoli, R., M. Laroche, M.A. Krieg, *et al.* 2010. Strontium ranelate and alendronate have differing effects on distal tibia bone microstructure in women with osteoporosis. *Rheumatol. Int.* **30:** 1341–1348.

66. Li, E.K., T.Y. Zhu, V.Y. Hung, *et al.* 2010. Ibandronate increases cortical bone density in patients with systemic lupus erythematosus on long-term glucocorticoid. *Arthritis Res. Ther.* **12:** R198.

67. Seeman, E., P.D. Delmas, D.A. Hanley, *et al.* 2010. Microarchitectural deterioration of cortical and trabecular bone: differing effects of denosumab and alendronate. *J. Bone Miner. Res.* **25:** 1886–1894.

68. Majumdar, S., D. Thomasson, A. Shimakawa, *et al.* 1991. Quantitation of the susceptibility difference between trabecular bone and bone marrow: experimental studies. *Magn. Reson. Med.* **22:** 111–127.

69. Weisskoff, R.M., C.S. Zuo, J.L. Boxerman, *et al.* 1994. Microscopic susceptibility variation and transverse relaxation: theory and experiment. *Magn. Reson. Med.* **31:** 601–610.

70. Ford, J.C., F.W. Wehrli & H.W. Chung. 1993. Magnetic field distribution in models of trabecular bone. *Magn. Reson. Med.* **30:** 373–379.

71. Link, T.M., S. Majumdar, P. Augat, *et al.* 1998. Proximal femur: assessment for osteoporosis with T2* decay characteristics at MR imaging. *Radiology* **209:** 531–536.

72. Wehrli, F.W., J.C. Ford & J.G. Haddad. 1995. Osteoporosis: clinical assessment with quantitative MR imaging in diagnosis. *Radiology* **196:** 631–641.

73. Krug, R., S. Banerjee, E.T. Han, *et al.* 2005. Feasibility of in vivo structural analysis of high-resolution magnetic resonance images of the proximal femur. *Osteoporos. Int.* **16:** 1307–1314.

74. Phan, C.M., M. Matsuura, J.S. Bauer, *et al.* 2006. Trabecular bone structure of the calcaneus: comparison of mr imaging at 3.0 and 1.5 t with micro-ct as the standard of reference. *Radiology* **239:** 488–496.

75. Laib, A., D.C. Newitt, Y. Lu, *et al.* 2002. New model-independent measures of trabecular bone structure applied to in vivo high-resolution mr images. *Osteoporos. Int.* **13:** 130–136.

76. Majumdar, S., D. Newitt, A. Mathur, *et al.* 1996. Magnetic resonance imaging of trabecular bone structure in the distal radius: relationship with x-ray tomographic microscopy and biomechanics. *Osteoporos. Int.* **6:** 376–385.

77. Hwang, S.N., F.W. Wehrli & J.L. Williams. 1997. Probability-based structural parameters from three-dimensional nuclear magnetic resonance images as predictors of trabecular bone strength. *Med. Phys.* **24:** 1255–1261.

78. Pothuaud, L., A. Laib, P. Levitz, *et al.* 2002. Three-dimensional-line skeleton graph analysis of high-resolution magnetic resonance images: a validation study from 34-microm-resolution microcomputed tomography. *J. Bone Miner. Res.* **17:** 1883–1895.

79. Majumdar, S., M. Kothari, P. Augat, *et al.* 1998. High-resolution magnetic resonance imaging: Three-dimensional trabecular bone architecture and biomechanical properties. *Bone* **22:** 445–454.

80. Ammann, P. & R. Rizzoli. 2003. Bone strength and its determinants. *Osteoporos. Int.* **14**(Suppl 3): S13–S18.

81. Link, T.M., J. Bauer, A. Kollstedt, *et al.* 2004. Trabecular bone structure of the distal radius, the calcaneus, and the spine: which site predicts fracture status of the spine best? *Invest. Radiol.* **39:** 487–497.

82. Majumdar, S., H. Genant, S. Grampp, *et al.* 1997. Correlation of trabecular bone structure with age, bone mineral density and osteoporotic status: in vivo studies in the distal radius using high resolution magnetic resonance imaging. *J. Bone Miner. Res.* **12:** 111–118.

83. Majumdar, S., T. Link, P. Augat, *et al.* 1999. Trabecular bone architecture in the distal radius using mr imaging in subjects with fractures of the proximal femur. *Osteoporos. Int.* **10:** 231–239.

84. Link, T.M., S. Majumdar, P. Augat, *et al.* 1998. In vivo high resolution mri of the calcaneus: differences in trabecular structure in osteoporosis patients. *J. Bone Miner. Res.* **13:** 1175–1182.

85. Benito, M., B. Gomberg, F.W. Wehrli, *et al.* 2003. Deterioration of trabecular architecture in hypogonadal men. *J. Clin. Endocrinol. Metab.* **88:** 1497–1502.

86. Benito, M., B. Vasilic, F.W. Wehrli, *et al.* 2005. Effect of testosterone replacement on trabecular architecture in hypogonadal men. *J. Bone Miner. Res.* **20:** 1785–1791.

87. Chesnut, C.H., 3rd, S. Majumdar, D.C. Newitt, *et al.* 2005. Effects of salmon calcitonin on trabecular microarchitecture as determined by magnetic resonance imaging: results from the quest study. *J. Bone Miner. Res.* **20:** 1548–1561.

88. Gomberg, B.R., P.K. Saha & F.W. Wehrli. 2005. Method for cortical bone structural analysis from magnetic resonance images. *Acad. Radiol.* **12:** 1320–1332.

89. Goldenstein, J., G. Kazakia & S. Majumdar. 2010. In vivo evaluation of the presence of bone marrow in cortical porosity in postmenopausal osteopenic women. *Ann. Biomed. Eng.* **38:** 235–246.

90. Wehrli, F.W. & M.A. Fernández-Seara. 2005. Nuclear magnetic resonance studies of bone water. *Ann. Biomed. Eng.* **33:** 79–86.

91. Krug, R., P.E.Z. Larson, C. Wang, *et al.* 2011. Ultrashort echo time MRI of cortical bone at 7 tesla field strength: a feasibility study. *J. Magn. Reson. Imaging.* doi: 10.1002/jmri.22648. [Epub ahead of print].

92. Techawiboonwong, A., H.K. Song, M.B. Leonard, *et al.* 2008. Cortical bone water: in vivo quantification with ultrashort echo-time MR imaging. *Radiology* **248:** 824–833.

93. Li, X., D. Kuo, A.L. Schaefer, *et al.* 2011. Quantification of vertebral bone marrow fat content using 3 Tesla MR spectroscopy: reproducibility, vertebral variation, and applications in osteoporosis. *J. Magn. Reson. Imaging* **33:** 974–979.

94. Meunier, P., J. Aaron, C. Edouard, *et al.* 1971. Osteoporosis and the replacement of cell populations of the marrow by adipose tissue. A quantitative study of 84 iliac bone biopsies. *Clin. Orthop. Relat. Res.* **80:** 147–154.

95. Justesen, J., K. Stenderup, E.N. Ebbesen, *et al.* 2001. Adipocyte tissue volume in bone marrow is increased with aging and in patients with osteoporosis. *Biogerontology* **2:** 165–171.

96. Bredella, M.A., P.K. Fazeli, K.K. Miller, *et al.* 2009. Increased bone marrow fat in anorexia nervosa. *J. Clin. Endocrinol. Metab.* **94:** 2129–2136.

97. Yeung, D.K., J.F. Griffith, G.E. Antonio, *et al.* 2005. Osteoporosis is associated with increased marrow fat content and decreased marrow fat unsaturation: a proton MR spectroscopy study. *J. Magn. Reson. Imaging* **22:** 279–285.